Reliability Based Aircraft Maintenance Optimization and Applications

Reliability Based Aircraft Maintenance Optimization and Applications

He Ren

Xi Chen

Yong Chen

Academic Press is an imprint of Elsevier
125 London Wall, London EC2Y 5AS, United Kingdom
525 B Street, Suite 1800, San Diego, CA 92101-4495, United States
50 Hampshire Street, 5th Floor, Cambridge, MA 02139, United States
The Boulevard, Langford Lane, Kidlington, Oxford OX5 1GB, United Kingdom

Library of Congress Cataloging-in-Publication Data
A catalog record for this book is available from the Library of Congress

British Library Cataloguing-in-Publication Data
A catalogue record for this book is available from the British Library

ISBN: 978-0-12-812668-4

For information on all Academic Press publications visit our website at
https://www.elsevier.com/books-and-journals

Publisher: Jonathan Simpson
Acquisition Editor: Glyn Jones
Editorial Project Manager: Naomi Robertson
Production Project Manager: Jason Mitchell
Designer: Mark Rogers

Typeset by Thomson Digital

Contents

List of Figures

List of Tables

About the Authors

Prof. He Ren is the acting chief engineer at the Shanghai Aircraft Customer Service Centre of Commercial Aircraft Corporation of China (COMAC). His task is science and technology innovations in ARJ21 and C919 programs. His research has covered the areas of aircraft reliability, safety, airworthiness certification, maintenance, and so forth.

Prof. Ren has extensive working experience globally. He has worked for aeronautical industries for more than 18 years, including aviation industries in China, Airbus, Boeing, Royal Air Force Australia (RAAF), and Defence Science and Technology Organization (DSTO–Australia). Furthermore, he has served as senior academic for universities such as Newcastle University, RMIT University (Australia), Nanjing Aeronautics and Astronautics University, and others for 12 years. Moreover, Prof. Ren plays an active role in professional societies. He is the fellow or a member in SAE, AIAA, the Engineer Institute of Australia, and Royal Aeronautical Society and was honorably awarded a number of guest professorships in universities. His effective and dedicated researches have resulted in about 10 PhD theses, and have been presented as keynote speeches at international conferences. He has received many awards for his contributions to technological progress and participated in aircraft projects such as JH7, MPC75, AE100, Y7-200B, J35, PAX750, A380, B707-Tank, ARJ21, and C919.

Dr. Xi Chen is a graduate PhD student of RMIT University and earned a postdoctoral fellowship in the Shanghai Aircraft Customer Service Centre of the Commercial Aircraft Corporation of China (COMAC).

Prof. Yong Chen is the chief designer of ARJ21 aircraft and the fellow member of the Science and Technology Committee of the Commercial Aircraft Corporation of China (COMAC).

Preface

Aircraft maintenance is an independent, multiskilled practice that synthesizes many disciplines from different sources, such as failure mode and effect analysis, damage and special events analysis, logistics-related operations analysis, software support analysis, and more. Therefore, reliable, systematic methodologies are needed for supporting the continuous airworthiness of newly developed regional jets or trunk airliners.

This book was written based on the researches of its authors, and initially used as an internal reference manual by aviation industries. Many thanks are given to Shanghai Jiaotong University Press and Elsevier Limited. It is their dedicated work that made these technical notes and papers possible for public available.

My sincere thanks are due to my research associates and fellows in the research team, Dr. Xi Chen, Dr. Supanee Arthasartsri, Mr. Ian Sutherland, Prof. Mingbo Tong, Prof. Yong Chen, and Prof. Hongfu Zuo. It is their expertise and assistance that made this book available.

Thanks are also given to my colleagues at the Commercial Aircraft Corporation of China (COMAC). Special thanks to Jin Zhuanglong, chairman of the board of COMAC; He Dongfeng, president of COMAC; and Xu Qinghong and Xu Jun, the president and vice president of the Customer Service Center of COMAC.

Finally, I thank my friends and previous colleagues for their assistance. They are Prof. Aleks Subic, the Deputy Vice Chancellor of Swinburne University; Prof. Chun Wang, the Head of the Mechanical School of New South Wales University; Prof. Adrian Mouritz, the Dean of Engineering and Associate Prof. Cees Bil of RMIT University of Australia.

And, of course, my deepest gratitude goes to my wife, daughters, brothers, and sisters for their eternal love and continuing support.

Prof. He Ren

Commercial Aircraft Corporation of China (COMAC), China

Abbreviations

AD Accidental damage
ADT Administration delay time
ANN Artificial neural network
APU Auxiliary power unit
ASD Aerospace and defense industries association of Europc
A-SHM Automated—structural health monitoring
ATA Air Transport Association
BBM Business-based maintenance
BCM Business-centered maintenance
BFRP Boron fiber reinforced plastic
BPN Back propagation network
BVID Barely visible impact damage
CAMM Computer-aided maintenance management
CBM Condition-based maintenance
CBT Computer-based training
CDF Cumulative distribution function
CE Concurrent engineering
CFRP Carbon fiber reinforced plastic
CMMS Computerized maintenance management system
COMAC Commercial Aircraft Corporation of China
D&SEA Damage and special events analysis
DI/DET Detailed inspection
DMC Direct maintenance cost
DMU Digital mock-up
DSI Dent spot index
ED Environmental deterioration
FAA Federal aviation administration
FBM Fault-based maintenance
FCS Flight control system
FD Fatigue damage
FMEA Failure mode and effect analysis
FST Functionally significant item
FTA Fault tree analysis
GFRP Glass fiber reinforced plastic
GVI General visual inspection
IBM Inspection-based maintenance
ILS Integrated logistic support
IWG Industry working group
KFRP Kevlar fiber reinforced plastic

LCC Life-cycle cost
LDT Logistic delay time
LORA Level of repair analysis
LROA Logistic-related operations analysis
LSA Logistics support analysis
LSAR Logistic support analysis record
MAARS Maintenance analysis and reporting system
MAMT Mean active maintenance time
MDT Maintenance down time
MEA Maintenance engineering analysis
MPD Maintenance planning document
MPMT Mean preventive maintenance time
MRBR Maintenance review board report
MRD Maintenance requirements document
MSG-3 Maintenance Steering Group-3
MSI Maintenance significant item
MTTF Mean time to failure
MTTR Mean time to repair
NDE Nondestructive evaluation
NDI Nondestructive inspection
OEM Original equipment manufacturers
PDF Probability density function
PMO Preventive maintenance optimization
POD Probability of detection
POF Probability of failure
PPH Policy and procedure handbook
R&M Reliability and maintainability
RAAF Royal Australian Air Force
RCM Reliability-centered maintenance
SDI Special detailed inspection
SHM Structural health monitoring
SMA Scheduled maintenance analysis
SRM Structural repair manual
S-SHM Scheduled-structural health monitoring
SSI Significant structural item
TBM Time-based maintenance
TMP Technical maintenance plan
TPM Total productive maintenance
V&V Validation and verification

Abstract

Aircraft maintenance is one of the critical operational tasks to sustain continued airworthiness. It also contributes a significant proportion of the total life-cycle cost of an aircraft. Based on the introduction to fundamental reliability theories, this book presents some solutions to the issues of integrated logistic and maintenance management. By overcoming the shortage of MSG-3 structural analysis in maintenance practices, this book offers flexible and cost-effective maintenance schedules for aircraft structures, particularly those with composite airframes. By applying an intelligent rating system, the back-propagation network (BPN) method, and FTA technique, a new approach is created with an acceptable learning curve and a flexible data fusion capability, to assist in determining inspection intervals for new aircraft structures. This book also discusses the influence of structure health monitoring (SHM) on scheduled maintenance. An integrated logic diagram was established, incorporating SHM into the current MSG-3 structural analysis, based on which four maintenance scenarios with gradual increasing maturity levels of SHM were analyzed. The inspection intervals and the repair thresholds are adjusted according to different combinations of SHM tasks and scheduled maintenance. This book provides a practical means for aircraft manufacturers and operators to consider the feasibility of SHM by examining labor work reduction and structural reliability variation, as well as maintenance cost savings.

This book can be used as a reference for aircraft designers, manufacturers, and operators, as well as serving as a textbook for students in educational institution.

Chapter 1

Introduction

Chapter Outline

1.1 CHALLENGES OF MODERN DEVELOPING COMMERCIAL AIRCRAFT

In the new millennium, economy and development have grown significantly to accommodate a rising number of air travelers. Future demands for increased convenience and safety in the aviation industry will be prompted by new developments in aircraft technologies. Driven by a strong economy, new entrants, large emerging markets, and increasing liberalization, air travel has grown nearly 30% since 2000, the strongest recovery in aviation history [1]. According to the forecast by Airbus, world passenger traffic is expected to increase by 4.8% per annum. In the largest emerging market, China, aviation passenger traffic volume grew 3.6 times, greater than the growth in railway and highway traffic volume during the period of 2001–12 [2]. It is estimated that more than 3000 new aircraft are needed in the next 20 years for the domestic market alone, and that by the year 2032 the volume of passenger traffic will account for 16% of the world's total, approaching the scale of the North America region [3].

To meet the booming civil aviation demand, new generation aircraft with modern technology is designed to be safer, more comfortable, and with greater fuel efficiency. The number and scale of the airport is being expanded to increase the capacity of airplane accommodation. Moreover, the operation efficiency is of key importance as the operational cost accounts for a large portion of the life-cycle cost. From the perspective of system engineering, a scientific maintenance strategy that is determined at the beginning or is updated in time

Reliability Based Aircraft Maintenance Optimization and Applications
http://dx.doi.org/10.1016/B978-0-12-812668-4.00001-0

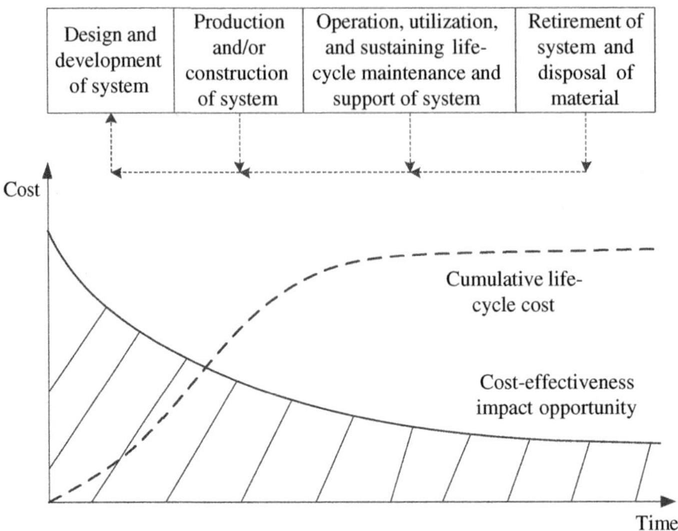

FIGURE 1.1 Opportunity for affecting logistics and system effectiveness.

can assure cost-effective aircraft operation and high flight safety. A typical figure indicating the cost relationship is shown in Fig. 1.1 [4].

Aircraft maintenance is developed to be an independent multidiscipline subject. For example, Maintenance Engineering Analysis (MEA) is carried out to synthesize many programs from different disciplines, such as failure mode and effect analysis, damage and special events analysis, logistic-related operations analysis, and software support analysis, and so on. Then a systematic analysis is conducted in order to make proper maintenance plans and activities.

A continuous airworthiness maintenance program is a compilation of the individual maintenance and inspection functions utilized by an operator to fulfill total maintenance needs. Authorization to use a continuous airworthiness maintenance program is documented and is approved by the Federal Aviation Administration (FAA). The basic elements of continuous airworthiness maintenance programs comprise aircraft inspection; scheduled maintenance; unscheduled maintenance; engine, propeller, and appliance repair and overhaul; structural inspection program or airframe overhaul; and required inspection items. A traditional deviation of maintenance activity is shown in Fig. 1.2 [5].

Following approval by the FAA, engineering provides the work package to the maintenance units and monitors standards. Engineering tasks include providing technical documentation, technical fleet management and planning, airworthiness control, schedule planning, reliability monitoring, quality assurance, and training. Engineering is much more than a "technical function." It is a "knowledge function" that works very closely with the maintenance function to optimize maintenance programs, increase fleet reliability, and facilitate flexible deployment.

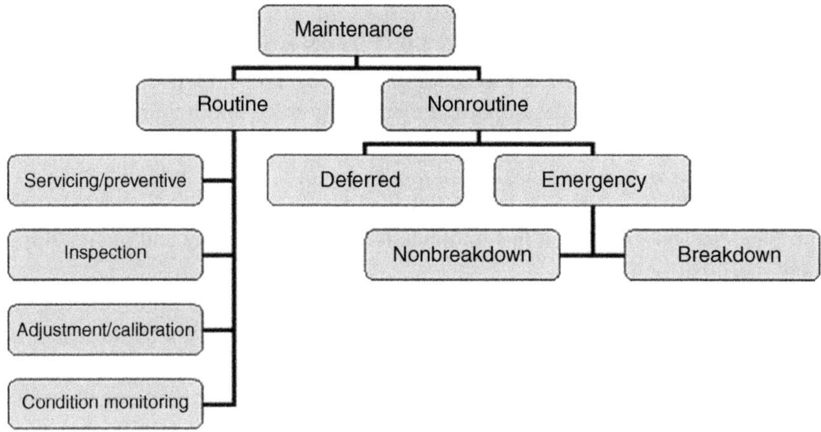

FIGURE 1.2 Types of maintenance activity.

With the fast development of modern technology, new materials and design concepts are being integrated into new aircraft and, thus, the traditional or existing maintenance programs may not be competent to the new requirements. For example, advanced sensors and data processing capability are promoting innovative monitoring methods, which may exert a profound influence on the current scheduled maintenance.

1.2 EVOLUTION OF AIRCRAFT MAINTENANCE PROCESS

The principle behind the construction of modern aircraft maintenance schedule is a documentation produced by Air Transport Association (ATA) maintenance steering group (MSG). The concept started in the 1960s by FAA on the first generation of wide body aircraft, that is, the Boeing 747, DC10, and L1011. Before the application of MSG Logic, hard time (HT) principal was in use, which based maintenance for the aircraft on the theory of preventive, yet expensive, replacement or restoration of components [6].

The process-oriented approach to maintenance uses three primary maintenance processes to accomplish the scheduled maintenance actions. These processes are called HT, on-condition (OC), and condition monitoring (CM) [7]. The HT and OC processes are used for components or systems that, respectively, have definite life limits or detectable wear-out periods. The CM process is used to monitor systems and components that cannot utilize either the HT or OC processes. These CM items are operated to failure, and failure rates are tracked to aid in failure prediction or failure prevention efforts. These are called "operate to failure" items.

The process used involved six industry working groups (IWGs), which includes structures, mechanical systems, engine and auxiliary power unit (APU), electrical and avionics systems, flight control and hydraulics, and zonal. Each group addresses their specific systems in the same way to develop an adequate initial maintenance program. The first MSG focuses on developing how to conduct a logical decision process to develop efficient, cost-effective maintenance routines that are acceptable to operators, manufacturers, and regulating authorities. The IWGs analyze each item using a logic tree to determine the requirements in the areas of functions, failure modes, failure effects, and failure causes. This approach to maintenance program development is called a "bottom up" approach because it looks at the components as the most likely causes of equipment malfunction [7].

Over time, the MSG process has evolved from a hard-time concept to CM. The process allows malfunctions to occur and relies upon the analysis of information about such malfunctions to determine the proper actions. To improve upon this method, MSG-2 was designed and then modified in 1980 in a document released by the ATA. Then, MSG-3 was built upon the existing framework of MSG-2. It adjusted the decision logic to provide a more straightforward and linear progression through the logic. MSG and MSG-2 are both bottom-up approaches; in contrast, the MSG-3 process is a top-down approach or consequence-of-failure approach. The component failures or deteriorations are not the main focus of the process; instead, the consequences of the failure and how it affects aircraft operations is considered. The idea is to cover and analyze each task based upon these three dimensions across the full decision tree. A simplified diagram [8] is shown in Fig. 1.3.

The result of the MSG-3 analysis constitutes the original maintenance program for the new model aircraft and the program that is to be used by a new operator of that model. The tasks selected in the MSG process are published by the

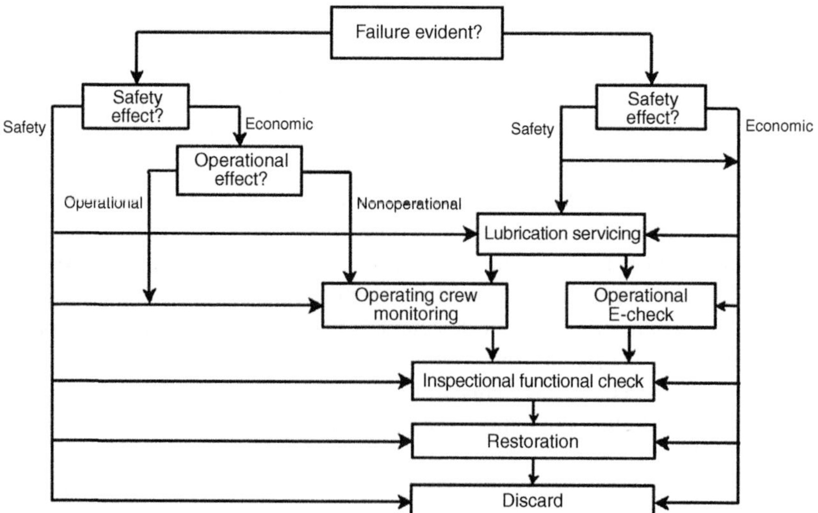

FIGURE 1.3 MSG-3 logic diagram.

airframe manufacturer in an FAA-approved document called the maintenance review board (MRB) report. This report contains the initial scheduled maintenance program and is used by those operators to establish their own FAA-approved maintenance program as identified by the operations specifications. The MRB report, the manufacturer publishes its own document for maintenance planning. For manufacturers like Airbus or Boeing, this document is called the maintenance planning document (MPD). This document often groups maintenance as an alphabetical checklist with hours, cycles, and calendar time. These estimated times must be altered by the operator to accommodate the actual task requirements when planning any given check activity.

1.3 AIRCRAFT COMPOSITE STRUCTURES

Composite materials are a new generation of materials that are increasingly used in the aviation industry. Since the 1970s, composite materials were first used on nonload bearing structures, such as radomes, fairings, and for inner decoration. In the 1980s, secondary structures began to be constructed with composite materials, but their use was still limited in structures like control surface panels. In the new millennium, the use of composite materials has shifted from secondary structures to primary structures. Typical examples are the world's largest aircraft, the Airbus 380, and the most advanced aircraft to date, the Boeing 787. More precisely, the composite structures used in the Airbus 380 weigh more than 30 tons, comprising 25% of the total airframe weight. The entire center wing box is made with composites [9]. The Boeing 787 adopts composite materials for the entire fuselage. Besides this, many components on the wing and nacelle are built with composite materials, so that composites account for 50% of the airplane [10]. Recently, the first prototype of the A350 was manufactured and the use of composite materials accounts for up to 52% of the plane [11], which marks very significant progress for Airbus and for the entire aviation industry. The development of composite materials by two leading aircraft manufacturers over the past two decades is depicted in Fig. 1.4.

Composite materials are formed by combining two or more constituent materials with significantly different physical or chemical properties to produce an integrated material with characteristics different from the individual ingredients [12]. The constituent materials have two main categories: matrix and reinforcement. The matrix materials surrounds and supports the reinforcement materials to maintain their relative positions. Meanwhile, the reinforcements provide special mechanical and physical properties to enhance the overall property. The wide variety of matrix and strengthening materials allows structure designers to optimize the combinations [13].

The matrices can be classified as metals or nonmetals. Aluminum, magnesium, titanium, and other metals are often used for a metallic matrix, while resins, ceramics, carbon, and so on are common materials for nonmetallic matrices. In terms of reinforced materials, carbon (including graphite), boron,

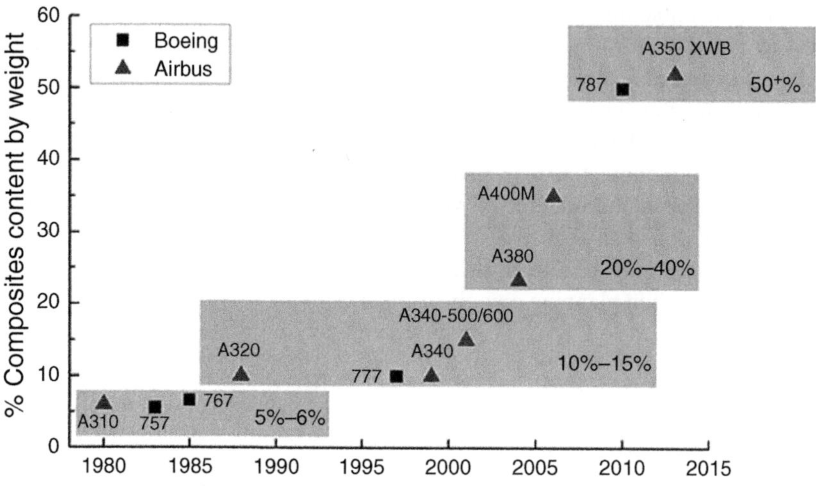

FIGURE 1.4 **Composite usage over last two decades.**

aramid, glass, and so forth form typical composites like carbon fiber reinforced plastic (CFRP), boron fiber reinforced plastic (BFRP), kevlar fiber reinforced plastic (KFRP), and glass fiber reinforced plastic (GFRP) [14].

Referring to the development of the current aviation industry, the most widely used composites in airframes are CFRP and the second is KFRP. Because of the high price and difficulty in fabricating BFRP, there is little use of BFRP. Compared to CFRP, GFRP has lower strength and stiffness properties and is generally not used for primary load-bearing structures, but since it is inexpensive, GFRP has applications in many secondary structures in civil aircraft.

There is another special class of hybrid composite material called sandwich. It is fabricated by attaching two thin metal skins to a lightweight but thick core. The hybrid composite structure has a high bending stiffness and also offers impact protection with overall low density.

The reason why composite materials have become attractive for the aviation industry is its unique properties, such as high specific strength and stiffness, fatigue resistance, long duration, and design adaptability to various loading conditions, etc. The most typical advantage of composites is the reduction of structural weight while maintaining the same loading capacity. This can lead to significant savings in life-cycle cost due to reduced fuel consumption. Moreover, due to advanced manufacturing processes, such as resin transfer molding (RTM), automated tape layup and automated fiber placement, the number of joints, and assembly parts can be greatly reduced through robust fabrication methods.

However, composite structures, compared to metallic structures, have more complex damage modes because of their anisotropic properties. One of the serious disadvantages is the susceptibility to impacts caused by runway debris, hail, tool dropping, and so on during operation. Object impact can cause internal damage, such as delamination or debonding, requiring intrusive inspections and repair activities [15]. Both Boeing and Airbus have placed significant emphasis on the use of composites in design and manufacturing as well as in maintenance during operation. The new aircraft developed in China will also use a large proportion of advanced composite materials. With accumulated experience in composite design and manufacturing, the maintenance of composite airframes becomes a big challenge.

1.4 RELIABILITY-CENTERED MAINTENANCE

1.4.1 Reliability Design

The reliability theory and relevant methodologies have been developed via several phases. There were three main technical areas evolved during the growth process:

1. Reliability engineering, which includes system reliability analysis, design review, and related task;
2. Operation analysis, which includes failure investigation and corrective action; and
3. Reliability mathematics, which includes statistics and related mathematical knowledge.

In order to achieve a better way to balance the cost of failure reduction against the value of the enhancement, accurately assessing failure rate of a system is necessary. The quantified reliability-assessment is one basic technique [16].

In the past, reliability measures centered on mechanical equipment and hardware. A reliable technology is the result of lessons learned from failure or experiment. The "test and correct" principle was used before formal data collection and analysis procedure development. During the design phase, to maximize reliability, the feedback principle was practiced through formal data collection techniques, which is very useful in improving inherent reliability. The failure data form the basis of reliability research. Failure data was manipulated and calculated to get the failure rate.

During 1940s the major statistical effect on reliability problem was in the area of quality control. As the equipment and systems becoming bigger, more complex and expensive, the traditional approaches become impractical in front of new complex objects. Very little experience could be gained from previous failure in most case since the extremely growth of complexity and cost of whole system of product, such as jet aircraft or nuclear power plant system.

Estimates of the reliability of equipment or complex system depend heavily on the field of mathematics known as statistics and probability. Even at a fairly elementary level, probability opens the door to the investigation of complex systems and situations. The language of probability is adapted to answer such questions as "What is the chance of that happening?" or "How much do we expect to gain if we make the decision?" However, it was not till the Korean War that quantitative reliability became widely used and statistics methods were applied to its measurement [17].

Weibull (1951) first proposed a distribution system, which was later named Weibull distribution. Squarely addressing the problem of tube reliability, the airlines set up an organization called Aeronautical Radio, Inc. (ARINC), which collected and analyzed defective tubes and returned them to the tube manufacturer. ARINC achieved significant success in improving the reliability of a number of tube types. The ARINC program has been focused on military reliability problems since 1950.

Three characteristic sorts of failures are identified as "early failure," "wear out failure," and "chance failure." These types of failure follow a specific statistic distribution and require different mathematical methods to treat, and different methods must be used for their estimates. For example, wear-out failures usually cluster around the mean wear-out life of components, so the probability of component wear-out failure occurrence at any given operation period can be mathematically calculated according to their failure distribution. Meanwhile, the early and chance failure usually occur at random intervals, they have characteristic distributions that are different from wear-out failure, and the probability of their occurrence in a given operation period can also be mathematically calculated.

In the mass production age, the cost of assuring reliability is high for the manufacturers. A balance is sought between reliability and benefits. This has led to the higher demand for quantifiable reliability-assessment techniques. Reliability prediction modeling techniques are produced by using the valid repeatable failure rate of a standard component to calculate and estimate the reliability of equipment or system. The development of computer technology makes it easier to sort the data and analyze the failure mode of the failure.

In the electric engineering area, redundancy system design and environmental screening/stress test techniques, Fault Tree Analysis (FTA), and Failure Mode Effect and Catastrophic Analysis (FMECA) techniques are widely applied. In structural engineering, the first-order reliability methods (FORM) and second order reliability methods (SORM) have evolved. However, the common weakness of conventional methods is failure to describe the nature of malfunction in a micro-process.

In addition, the Failure Mode and Effect Analysis (FMEA) is an intelligent response surface method based on a simplified model; it is a successful tool for system reliability analysis. Monte Carlo simulation (MCS) is a versatile tool to analyze and estimate the reliability and maintainability of a complex system. FTA is another widely used tool for system risk assessment. The FTA, using fuzzy failure probability, has the following advantages: it is not necessary to determine crisp values of the failure and error probabilities of basic events in a fault tree [18]. Fuzzy theory can be a useful tool to complement probability theory.

Nowadays, the subject of reliability prediction, based on the concept of validly repeatable component failure rates, has become controversial. The failure rates of complex products or systems do not always result from component failures, which are usually identified under approximately identical environmental and operating conditions. The factors influencing the reliability of a complex system are widely various: they could include software elements, human factors or operating documentation, or even continuously changing environmental factors. The system reliability model and the relationship among contribution factors are also becoming more complicated. The hypotheses of conventional reliability theory are also the limitation of their application. So the theory and methodologies with the fuzzy set and MCS have been developed to supplement the conventional reliability theory.

1.4.2 Reliability-Centered Maintenance

Reliability-Centered Maintenance, often known as RCM, is an industrial improvement approach focused on identifying and establishing the operational, maintenance, and capital improvement policies that will manage the risks of equipment failure most effectively. It is defined by the technical standard SAE JA1011, *Evaluation Criteria for RCM Processes*. Nowlan and Heap [19] defined RCM as *a scheduled maintenance program designed to realize the inherent reliability capabilities of equipment*. Moubray [20] defined this as *a process used to determine what must be done to ensure that any physical asset continues to do what its users want it to do in its present context*. Another definition is *a systematic consideration of system functions, the way function can fail, and a priority-based consideration of safety and economics that identifies applicable and effective PM (preventive maintenance) tasks* [21].

The objective of RCM program is to realize the inherent reliability capabilities of the equipment for which they are designed, and to do so at minimum cost. Rausand [21] also suggested that it is to reduce the maintenance cost, by focusing on the most important functions of the system, and avoiding or removing maintenance actions that are not strictly necessary. If a maintenance program already exists, the result of an RCM analysis will often be to eliminate inefficient PM tasks. The principles of RCM stem from a rigorous examination of certain questions that are often taken for granted. How does a failure occur? What are its consequences? What good can preventive maintenance do?

It is agreed that maintenance is primarily aimed at preserving system function, not preserving equipment. Therefore, it is essential to understand what the expected outcome should be, and that the primary task is preserving that outcome or function [22]. There are four features that characterize RCM: (1) preserve system function, (2) identify failure modes that can defeat the functions, (3) prioritize the function need (via the failure modes), and (4) select only applicable and effective PM tasks.

The RCM application was introduced to the aviation industry in 1974 by the United States Department of Defense and United Airlines. It is also known as the MSG-2. Driven by accumulated experience over several years' use, an update was conducted and a methodology for designing maintenance programs based on tests and proven airline practices was documented by ATA, which formed the basis for MSG-3. The maintenance guidance MSG-3 remains to this day the process used to develop and refine maintenance programs for all major types of civil aircraft.

SAE JA1011 stated seven basic questions for RCM as follows:

1. What are the functions and associated performance standards of the asset in its present operating context?
2. In what ways does it fail to fulfill its functions?
3. What causes each functional failure?
4. What happens when each failure occurs?
5. In what way does each failure matter?
6. What can be done to predict or prevent each failure?
7. What should be done if a suitable proactive task cannot be found?

The steps of RCM analysis in order to answer the seven questions above are shown in Fig. 1.5.

In terms of RCM implementation, industries, such as aircraft, offshore oil, nuclear power, and so on have successful application experiences. RCM offers significant improvement in system reliability and availability, while it also helps to increase safety and reduce the amount of preventive maintenance activities.

However, in other industries, such as power, processing, and manufacturing, RCM is mainly applied to preexisting plants that are individually designed to meet a wide range of output requirements. Another condition is the available resources, which are usually established by custom and usage, and the time of introduction of RCM in terms of restraint and rationalization [23]. The

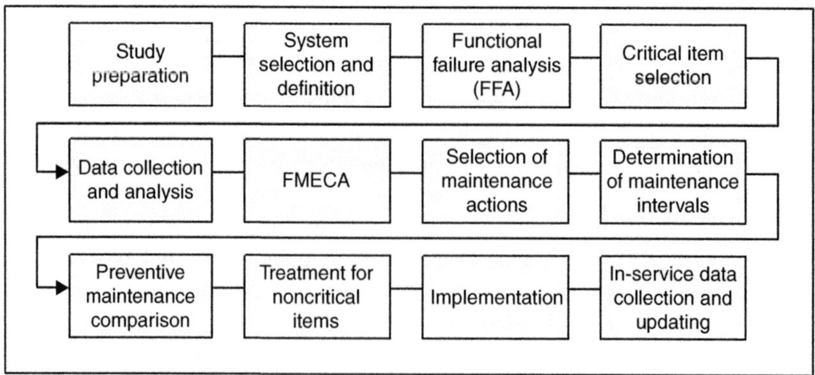

FIGURE 1.5 Main RCM analysis steps.

problems and deficiencies that became obstacles to the progress of the RCM are listed as follows:

Lack of a computerized maintenance management system (CMMS). This makes it difficult to gather and handle the information and data needed to support initial RCM analyses and to make optimizations.
Lack of an RCM computer system, which is required to handle the many analyses that may be made.
Lack of plant register, which makes it difficult to develop the RCM computer system due to lack of a system structure that is valid for all plants in the corporation. This also makes it difficult to gather information.
Unavailability of documentation and information, which makes it difficult for RCM teams to make correct analyses, and the lack of historical reliability data, which makes it difficult to conduct probability assessment.
Problematic routines, roles, and responsibilities, which lead to technical staff and maintenance personnel not being involved in the introduction. This makes it difficult to approve recommendations in a timely fashion.
Communication problems among technical staff, middle management, operators, and maintenance personnel, regarding the meaning and application of the recommendations arising from the RCM analyses.
Lack of an overarching maintenance management strategy, which makes it difficult to determine how to handle lists and plans pertaining to maintenance activities.
Incomplete goal setting and benefit identification and measurement. The criteria on which results should be based were to some extent unclear.

Regardless of the fact that RCM is today the most accepted and broadly used strategy in industry, other maintenance strategies have been developed, including:

Preventive Maintenance Optimization (PMO): A strategy originally described by Steve Turner (2001) and aimed at continuously reviewing and updating the maintenance program based on failure history, changing operating circumstances, and the advent of new predictive maintenance technologies.
Total Productive Maintenance (TPM): This strategy is an innovative Japanese concept, a maintenance program that involves a newly defined concept for maintaining plants and equipment. The goal of the TPM program is to markedly increase production while, at the same time, increasing employee morale and job satisfaction.
Business Centered Maintenance (BCM): BCM is an attitude, concept, and process of continuous improvement in maintenance and maintenance processes, equipment condition, and performance to improve overall equipment effectiveness, operations efficiency, output quality, and worker safety. It targets results by using a common sense approach that recognizes that

maintenance, production, and engineering are a partnership—engaged in a joint venture to produce quality products at the lowest cost.

Business Based Maintenance (BBM): This strategy was developed by Siemens and originally based on BCM, but is specifically for emergency maintenance. The method is to determine production processes requirements jointly with maintenance tasks and activities.

Computerized Maintenance Management Systems: CMMS assist in managing a wide range of information on maintenance workforce, spare-parts inventories, repair schedules, and equipment histories. It may be used to plan and schedule work orders, to expedite dispatch of breakdown calls and to manage the overall maintenance workload. CMMS can also be used to automate the PM function, and to assist in the control of maintenance inventories and the purchase of materials. CMMS has the potential to strengthen reporting and analysis capabilities.

Others: Some systems or components do not actually require any of the above strategies, and instead can be fixed when broken (run-to-failure), or replaced at the end of the life of the component (life-cycle method), or given zero risk–based maintenance, and so on.

1.5 MSG-3 STRUCTURAL ANALYSIS

MSG-3 presents a means for developing scheduled maintenance tasks and intervals that are intended to be accepted by the regulatory authorities, the operators, and the manufacturers [8]. The primary objective of the scheduled structural maintenance program is to maintain inherent airworthiness throughout the operational life of the aircraft in an economical manner. It is inherited and evolved from Reliability-Centered Maintenance (RCM) concept and can be deemed as a civil aircraft version. The result of MSG-3 analysis constitutes the original maintenance programs for the new type of aircraft. An approved document containing the selected tasks in the MSG-3 process by the aircraft manufacturer is called Maintenance Review Board Report (MRBR). This report specifies the initial scheduled maintenance programs and another document made by the manufacturer called MPD is developed based on MRBR. MPD is then used by the operators to establish their own regulatory approved maintenance programs to accommodate their practical situations.

The MSG-3 analysis begins with the development of a complete breakdown of the aircraft systems, down to the component level. All structural items are identified as either structure significant items (SSIs) or other items. MSG-3 defines what SSIs contribute significantly to carrying flight, ground, pressure or control loads, and which component failures could affect the structural integrity necessary for the safety of the aircraft. SSIs are then listed and categorized as damage-tolerant or safe life items, in which case life limits are assigned.

The assessment of SSIs for the selection of maintenance task should consider the following three damage sources:

1. *Accidental Damage (AD)*: Sources of such damage include ground and cargo handling equipment, foreign objects, erosion from rain, hail, lightning, runway debris, spillage, freezing, thawing, and so forth, and damage resulting from human error during aircraft manufacture, operation, or maintenance.
2. *Environmental Deterioration (ED)*, which is characterized by structural deterioration as a result of a chemical interaction with the climate or environment. Assessments are required to cover corrosion, including stress corrosion, and deterioration of nonmetallic materials.
3. *Fatigue Damage (FD)*, which is characterized by the initiation of a crack or cracks due to cyclic loading and subsequent propagation. It is a cumulative process with respect to aircraft usage (flight cycles or flight hours).

The assessment for selecting tasks and intervals also includes the analysis of the susceptibility of the structure to each source of deterioration, the consequences to airworthiness analysis, and the applicability and effectiveness of various methods of preventing, controlling, or detecting the deterioration, taking into account inspection thresholds and repeat intervals [8]. The process is summarized and illustrated in the following diagram, Fig. 1.6.

The structural maintenance program has been developed by considering each source of damage as stated in MSG-3 guidelines.

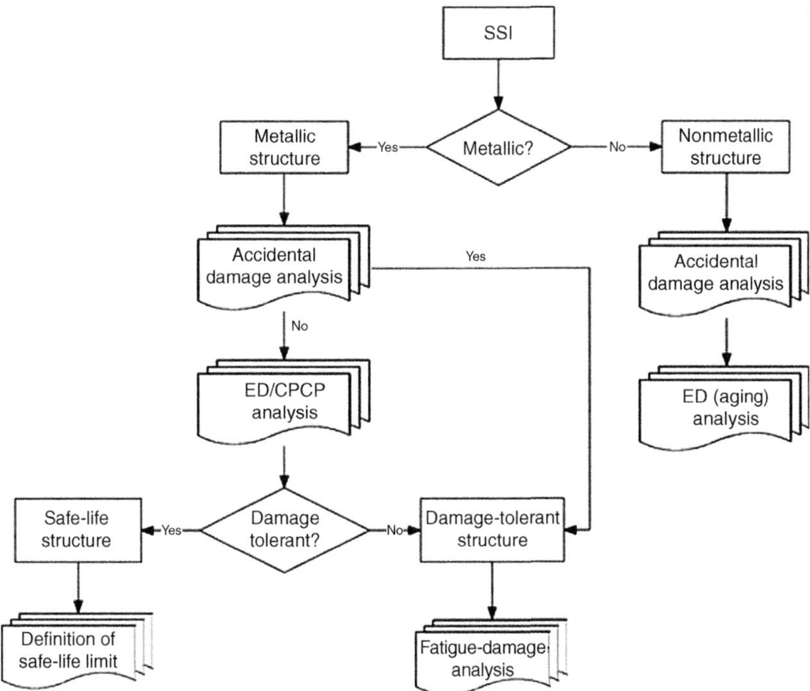

FIGURE 1.6 Simplified MSG-3 aircraft structure decision logic.

1. *Accidental Damage (AD)*: AD consists of minor damage that could result in fatigue and/or corrosion damage, and which could propagate undetected. In the analysis, the fatigue and environmental effects of the accidental damage are considered. A dedicated AD task is not selected; the AD requirement is consolidated into either the ED or the FD task, or both, according to the AD consequences. AD is therefore subject to maintenance requirements expressed in Flight Cycle FC (and Flight Hour FH) or calendar time or both.
2. *Environmental Deterioration (ED)*: For ED, since deterioration caused by the environment (e.g., corrosion, stress corrosion) is mainly time dependent, the maintenance requirements are based on calendar time (years).
3. *Fatigue Damage (FD)*: The damage initiation and subsequent damage growth is primarily dependent on the ground-air-ground loading variation, which occurs once per flight. Therefore, flight cycles (FC) are used as the unit for thresholds and repeat intervals. Some fatigue-related SSIs are also sensitive to flight duration. For these SSIs, a FH limit is stated in addition to the FC limit. The inspection is to be performed at whichever limit is reached first. The structure inspection tasks selected for fatigue are derived from damage tolerance evaluation according to the criteria defined by FAR/JAR 25.571, Amendment 45. The FC and FH data are based on a combination of calculation, full-scale fatigue test teardown results, and in-service experience where available.

Generally, there are three inspection levels, offering choices on which method is the most appropriate for the type of damage that is expected:

1. *General Visual Inspection (GVI)*: A visual examination of an interior or exterior area, installation or assembly to detect obvious damage, failure, or irregularity. This level of inspection is made from within touching distance unless otherwise specified. A mirror may be necessary to ensure visual access to all surfaces in the inspection area. This level of inspection is made under normally available lighting conditions, such as daylight, hangar lighting, flashlight or drop-light, and may require removal or opening of access panels or doors. Stands, ladders, or platforms may be required to gain proximity to the area being checked.
2. *Detailed Inspection (DET)*: An intensive visual examination of a specific structural area, system, installation, or assembly to detect damage, failure, or irregularity. Available lighting is normally supplemented with a direct source of good lighting at an intensity deemed appropriate. Inspection aids, such as mirrors, magnifying lenses, and so forth may be necessary. Surface cleaning and elaborate access procedures may be required.
3. *Special Detailed Inspection (SDI)*: An intensive examination of a specific item, installation or assembly to detect damage, failure, or irregularity. The examination is likely to make extensive use of specialized inspection techniques and/or equipment. Intricate cleaning and substantial access or disassembly procedure may be required.

Corrosion found during first or subsequent inspections, which is determined (normally by the operator) to be an urgent airworthiness concern requires expeditious action.

Corrosion Prevention and Control Task: A corrosion prevention and control task usually consists of: (1) removing equipment and interior furnishings to allow access to the area, (2) cleaning the area, (3) conducting inspections of all areas (Note: nondestructive inspections or visual inspections may be necessary), (4) removing all corrosion, evaluating damage, and repairing damaged structure, (5) unblocking drainage holes, (6) applying corrosion preventive compounds, and (7) reinstalling dry insulation blankets. An Implementation Threshold for a given area is the airplane age at which the CPCP should be implemented in the affected area. A Repeat Interval is the calendar time period between successive corrosion task accomplishments.

1.6 A380 MAINTENANCE PROGRAMS

In 2005, Airbus was building a new support strategy where customers pay for a significant portion of purchased services with data collected during operations. In a press briefing in December 2005 at the aircraft manufacturer's headquarters in Toulouse, France, executives from Airbus customer services team explained that rather than becoming a standalone business unit, integrated customer support could help make Airbus airplanes more attractive. In addition, they outlined plans to create a network of what MRO provides, and gave an extensive description of Airbus "e-solutions" for maintenance.

Because most airlines are outsourcing more maintenance and engineering tasks, many airlines have passed the nonreturn point. However, airlines are still seeking the right balance and solutions for maintenance planning. Paradoxically, Airbus support people are concerned that engineering tasks are being transferred to aircraft manufacturers. Mr. Gavin, an Airbus customer service executive, added that "diminishing engineering resources at the airlines will impact dispatch reliability" [24].

Airbus also has launched "a major improvement program." It ranges from "lease-loan tools" schemes to the aircraft maintenance analysis software tool (AIRMAN), plus flight operations monitoring, customized spares logistics, and A380 services. On the latter, Mr. Gavin noted that Airbus wants to take into account specific types of operations, both long haul and with a high number of passengers. "This implies that our system including Airbus and its vendors, such as engine manufacturers, has to be very reactive." Roger Leconte, Senior Vice-President for technical support and programs, said that product improvements and services to airlines should optimize maintenance and operational costs, respectively.

In brief, Airbus maintenance strategy concentrates on the service after the EIS to support airlines' MRO. The network is to be established to optimize the aftermarket service for the Airbus A380 fleet. This network is also a reaction to the general clamor by airlines for reducing maintenance costs. With new technology aircraft like A380, the most important thing is that there must a more complete customer support network in place than is usually required.

Airbus has set 99% dispatch reliability for the A380 within two years of EIS and 24% lower Direct Maintenance Cost (DMC) per seat than the Boeing 747-400 [25]. To achieve 99% reliability means only 1% can be allocated to failures across all the systems on the aircraft, including the structure, avionics, cabin systems, propulsion, and so on. The question is, What reliability and maintainability (R&M) strategy can Airbus use to design the A380 in order to achieve 99% dispatch reliability?

Airbus has developed new processes for a more rigorous evaluation of the design against the targets and they used two independent evaluation methods in the development. The first method uses an extensive availability model based on forecasting system malfunctions, which lead to potential delays or cancellations, and rectification times. The second method is based on the System Safety Assessment approach, where probabilities are assigned to potential failure conditions likely to result in flight or ground interruptions.

Designing the A380 for Operational Reliability (OR) followed a Verification and Validation (V&V) process, as shown in Fig. 1.7, in which the OR targets and requirements were defined at aircraft level and then broken down to system and equipment level validation. Then, verification, from equipment level up to aircraft level, was performed using simulation tools and test beds throughout the aircraft development and builds process.

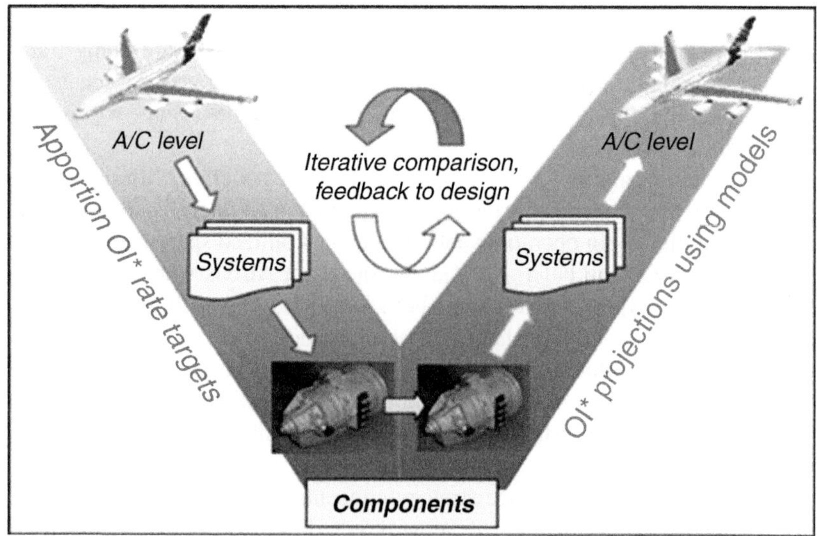

FIGURE 1.7 A380 verification and validation process.

During the manufacturing process, there is a far greater amount of testing than has even been done before prior to EIS. The A380 went through numerous testing with full-scale test rigs and a series of comprehensive flight tests. These test rigs include cabin zero, with an associated Internal Field Emission test-rig in Hamburg, landing gear zero in Filton, UK, and the iron bird (for the actuation and hydraulic systems) in Toulouse, a fuel test-rig.

To support reliability in service, airlines are demanding much better monitoring of the aircraft systems. To date, Airbus has offered systems monitoring as an option, but on the A380, it is a set-in-stone requirement. As a result, the aircraft features a lot of additional health monitoring sensors and software to observe every system in greater detail. Initial demands focused on monitoring the engines and auxiliary power unit (APU), but have since expanded to include other systems [26].

Alongside better systems monitoring, the A380 offers three further advances in troubleshooting technology. These are improved built-in test equipment (BITE), automated fault-reporting through ACARS satellite data links, and free online access to Airbus own troubleshooting software, AIRMAN 2000. The BITE is not same as the current other commercial aircraft with interactive BITE, as the engineer simply presses a button to run a test and the aircraft configures the system for the requested test, with no paper documentation required. All required manuals are stored and interlinked in the central maintenance system (CMS) while online links to AIRMAN 200 are available through ACARS.

Advanced monitoring systems and troubleshooting technologies are to support the reliability of the A380 in service. Time and practice become the key points for which the airline operators can accept the "new way of doing things" in order to reduce the maintenance costs on the aircraft.

In the maintainability aspect, Airbus has advanced maintainability optimization. The Original Equipment Manufacturers (OEM) devised a sophisticated virtual mannequin to conduct maintenance tasks in its aircraft digital mock-up. This mannequin was designed to make sure that maintenance personnel were able to do the work without injury or operating outside occupational health and safety regulations and limits. Such was its sensitivity that any abnormal stresses in the back, arms, or legs while lifting, moving or carrying equipment would be instantly highlighted. As a result, access to equipment and removal procedures have been simulated and optimized in order to minimize the downtime and maintain OR level.

1.7 SUMMARY

Aircraft maintenance is one of the critical operational tasks to sustain continued airworthiness. It also contributes a significant proportion of the total life-cycle cost. Based on introducing the fundamental concepts and theories of reliability and maintainability, some maintenance control and management methods are presented in this book. For overcoming the MSG-3 shortage in practice, this book is going to determine flexible and cost-effective maintenance schedule for aircraft structures particular in composite airframes. By applying an intelligent rating system, the back-propagation network (BPN) method, and FTA technique, a new approach was created with a powerful learning ability and a flexible data fusion capability, to assist in determining inspection intervals for new aircraft structures, especially in composite structure. Also, this book discusses the influence of Structure Health Monitoring (SHM) on scheduled maintenance. An integrated logic diagram was established incorporating SHM into the current MSG-3 structural analysis, based on which four maintenance scenarios with gradual increasing maturity levels of SHM were analyzed. The inspection intervals and the repair thresholds are adjusted according to different combinations of SHM tasks and scheduled maintenance. This book provides a practical means for aircraft manufacturers and operators to consider the feasibility of SHM by examining labor work reduction, structural reliability variation as well as maintenance cost savings. Finally, A380 Reliability and Maintainability program, as an example, is explained in this book.

Chapter 2

Basic Concepts

Chapter Outline

2.1 ACCIDENT

An *accident* is an undesirable event that may lead to loss of human life, personal injury, significant damage to the environment, or significant economic loss. The definition of accident is shown in Fig. 2.1.

2.1.1 Accident in Aviation

Accident, as defined by International Civil Aviation Organization (ICAO) Annex 13, means an occurrence associated with the operation of an aircraft in which any person suffers *fatal injury* or *serious injury*, in which the aircraft

Reliability Based Aircraft Maintenance Optimization and Applications
http://dx.doi.org/10.1016/B978-0-12-812668-4.00002-2

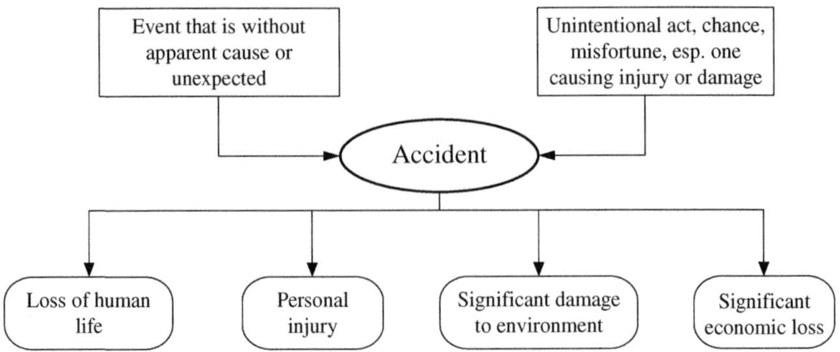

FIGURE 2.1 Accident definition.

receives *substantial damage*, or in which the aircraft is missing (an aircraft is considered to be missing when the official search has been terminated and the wreckage has not been located) or is completely inaccessible.

1. *Fatal accident*: an accident that result in one or more fatal injuries
2. *Fatal injury*: an injury resulting in death within 30 days of the date of the accident
3. *Serious injury* means any injury that:
   a. Requires hospitalization for more than 48 h, commencing within 7 days from the date the injury was received.
   b. Results in a fracture of any bone (except simple fractures of fingers, toes, or nose).
   c. Causes severe hemorrhages (or) nerve, muscle, or tendon damage.
   d. Involves any internal organ; or
   e. Involves second- or third-degree burns, or any burns affecting more than 5% of the body surface.
4. *Substantial damage* means damage or failure that adversely affects the structural strength, performance, or flight characteristics of the aircraft, and which would normally require major repair or replacement of the affected component. Substantial damage excludes damage to landing gear, wheel, tire, and flaps. It excludes bent aerodynamic fairings, dent in aircraft skin, small punctures in the aircraft skin, ground damage to propeller blades, or damage to only a single engine.

2.1.2 Accident Category in Aviation

- *Major*: an accident in which aircraft was destroyed, there were multiple fatalities, or there was one fatality and substantial damage to the aircraft.
- *Serious*: an accident in which there was either one fatality without substantial damage to the aircraft or at least one serious injury and any kind of substantial damage to the aircraft.

- *Injury*: a nonfatal accident with at least one serious injury and without substantial damage to the aircraft.
- *Damage*: an accident in which no person was killed or seriously injured, but in which any kind of substantial damage was incurred by the aircraft.

2.2 NEAR MISSES

A *near miss* is an undesirable event without loss of life and personal injuries, insignificant damage to the environment, and insignificant economic loss, but which, with small changes in the situation or in the state of the system, might have resulted in an accident. An unignited hydrocarbon leakage in a process plant will normally be categorized as a near miss. An ignited leakage will probably be identified as an accident.

2.3 RISK

Risk is defined as the danger that undesirable events present to human beings, the environment, and economic value. Risk can be expressed quantitatively in different ways, but usually by means of the frequency (probability) and the consequence of undesirable events, shown in Fig. 2.2.

In situations where the risk is related to loss of life, the so-called FAR Value (fatal accident rate) is often used to measure the risk level. The FAR is defined as the statistically expected number of accidental deaths per 100 million (10^8) exposed hours. At the time the FAR value was introduced, 10^8 h corresponded to the time 1000 people spent at their place of work over a lifetime. Today it takes 1400 people to reach 100 million working hours.

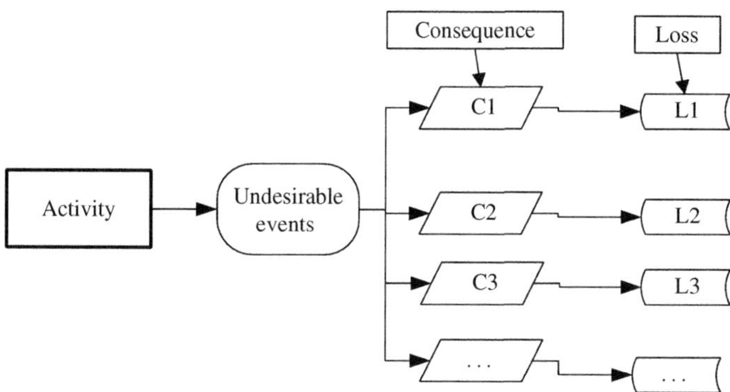

FIGURE 2.2 General risk model.

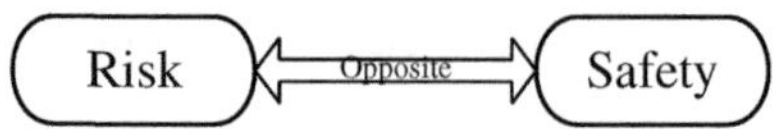

FIGURE 2.3 Safety.

2.4 SAFETY

Safety is characterized as the ability to avoid damage and loss as a consequence of undesirable events. The damage and loss can be related to effects on the lives and health of human beings, the biological and physical environment, or economic value (Fig. 2.3).

2.5 RELIABILITY

A few definitions of *reliability* and their relevance are shown in Fig. 2.4:

- A characteristic of the ability (probability) of a component or a system to perform its intended function for a specified time interval under stated conditions.
- Reliability is the quality over the time.
- Reliability is the probability of a product performing its intended function for a specified life under the operating conditions encountered in a manner that meets or exceeds customer expectations.

Normally, a reliability index could be denoted by:

- MTBF (mean time between faults)
- MTTF (mean time to failures)
- MTTR (mean time to repairs)
- Failure rate
- Failure probability

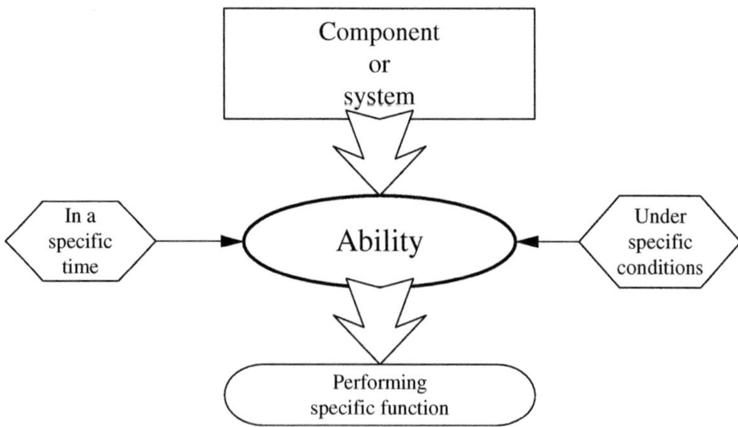

FIGURE 2.4 Reliability.

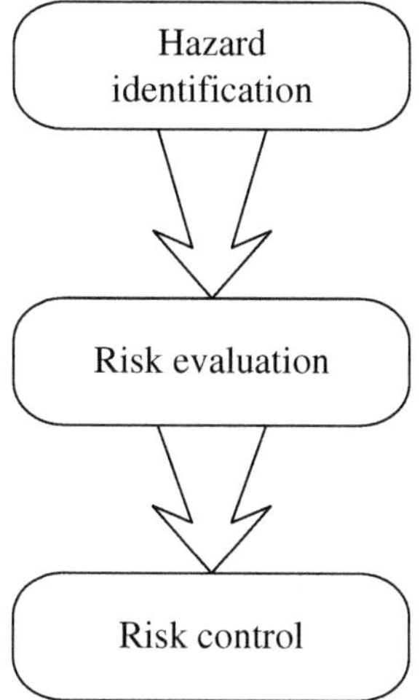

FIGURE 2.5 Risk management.

- Departure reliability
- Flight reliability

2.6 RISK MANAGEMENT

Risk management comprises three components: (1) risk identification, (2) risk evaluation, and (3) risk control, as shown in Fig. 2.5.

2.7 INCIDENT

An *incident* is an occurrence other than an accident, associated with the operation of an aircraft, that affects or could affect the safety of operations.

2.8 AIRWORTHINESS

Airworthiness is the certification that an air carrier has met the required standard for safety and operation and is authorized to provide aviation service.

2.9 QUALITY

Some definitions of *quality* are:

- *Customers define quality*: Customers want products and services that throughout their life meet and exceed their needs and expectations, at a cost that represents value.
- *ISO definition*: Quality is the totality of features and characteristics of product or service that bears on its ability to satisfy given needs.

Some characteristics of quality are price, safety, maintenance and service, and aesthetic value. Customers are highly influenced by following factors:

- *Performance*: refers to the primary operating characteristics of a product
- *Feature*: secondary characteristics that supplement the product's basic functioning
- *Reliability*: probability of a product failing within a specified period of time
- *Conformance*: degree to which a product's design and operating characteristics match preestablished standards
- *Durability*: a measure of product life, in both economic and technical dimensions
- *Serviceability*: speed, courtesy, and competence of repair
- *Aesthetics*: (subjective dimension) refers to how a product looks, feels, sounds, tastes, or smells
- *Perceived quality*: (subjective dimension) refers to the assessment of standards relying on indirect measure when comparing product brands

2.10 AIRWORTHINESS

The commonly accepted definition of *airworthiness* is the condition of an item (aircraft, aircraft system, or part) in which that item operates in a safe manner to accomplish its intended function.

A more explanatory definition covers the factors involved: "Airworthiness is a concept, the application of which defines the condition of an aircraft and supplies the basis for judgment of the suitability for flight, in that it has been designed, constructed, maintained, and is expected to be operated to approved standard limitations, by competent and approved individuals, who are acting as members of approved organization and whose work is both certified as correct and accepted on the operator."

2.11 AVAILABILITY

The term *availability* has significance in the measurement of maintenance performance, especially for military purposes, known as "operational readiness." This is defined as the proportion of time an aircraft is available to carry out its designated function. The level of availability achievable depends on a range of factors including inherent reliability of equipment, availability of spare parts and manpower, and the scheduled and unscheduled maintenance that must be performed. Blanchard [4] notes that availability is a function of operating time (reliability) and downtime (maintainability/supportability).

2.12 AIRCRAFT MAINTENANCE

The Services define *aircraft maintenance* as all actions taken to retain material in order to restore it to a specified condition or to restore it to serviceability.

A similar airline definition is "The provision of serviceable aircraft on time lines required by the operator" (SAE ARP 4741).

The apparent slight difference in the airline's emphasis on the operator's requirements is in fact more apparent than real. All aircraft operators civil or military have very similar objectives of safety, economy, and operational effectiveness in meeting the organisation's mission.

The maintenance function generally relies on equipment that has already been certified to an approved airworthiness standard. Maintenance preserves it in an operational condition, sustaining the original configuration and technical integrity performance requirements. The objective of maintenance activity is to preserve inherent levels of equipment safety and mission reliability and to achieve the required level equipment availability while utilizing available resources most efficiently.

The tasks or activities to be managed include:

- Servicing—which can include inspection
- Overhaul
- Bay servicing
- Repair and modification
- Replacement/throwaway
- Functional testing
- Calibration
- Nondestructive evaluation/inspection

The forms of maintenance are significantly distinguished as the following three categories:

1. Preventative maintenance
2. Surveillance maintenance
3. Corrective maintenance

A maintenance activity level is often used to identify the depth and complexity of a task. In most organizations there are three main levels of maintenance:

1. The operating level of maintenance is also called line maintenance. This refers to maintenance carried out in relation to a forthcoming flight and to subsequent minor repairs within the capabilities of line staff.
2. The heavy maintenance level means deeper maintenance services. It covers more extensive maintenance tasks requiring hangar facilities and a wide range of special equipment.
3. Intermediate level maintenance means maintenance activities on aircraft that, while not requiring the full facilities and skills of the deeper level, are done away from the flight line, in some form of centralized maintenance facility. Improvements in reliability, in basic technologies and maintenance management practices have generally made this level obsolete.

When dealing with aircraft maintenance, in most cases we are concerned with the systemthat supports aircraft operations and includes the following elements:

- Prime aircraft equipment
- Data—includes manuals, schedules, design drawings, parts lists
- Software
- Support equipment
- Spare parts
- Training aids and data
- Facilities
- Personnel
- Services—includes technical, supply, and personnel services

2.13 SOURCES AND TYPES OF FAILURE IN AIRCRAFT

What is the basis for our concern for serviceability and why does maintenance form such an important link in the system that aims to provide aircraft availability for operational missions? The risk of failure provides the basic rationale for maintenance. A failure in an aircraft can affect safety and mission achievement, cause secondary damage, and incur an economic penalty.

If we had a perfect machine, perfectly designed, manufactured flawlessly to a state from which it will not degrade, and if it were always operated within the design envelope, there would be little needed other than replenishment servicing. Because these requirements are not always met, and because the outcome of a failure with aircraft can be so catastrophic, there has to be surveillance, preventative maintenance, and corrective rectification and repair available to restore the imperfect machine to an appropriate level of ability. A few examples of types of failure and deterioration that can make an aircraft unserviceable will be reviewed in the following sections.

2.13.1 Mechanisms of Failure

In the management of maintenance one becomes very conscious of the failure mechanisms in aircraft and systems equipment around which maintenance is structured. Some of these are:

Material failure: A part may fail from many causes. Structural parts may crack or distort from corrosion, fatigue, fretting, external damage, overheating, and overload. Engine parts may be similarly affected by these processes as well as stress-generated fatigue, high temperature creep, erosion failures, and foreign object damage including bird strikes. Failures of plugs, micro switch contacts or wiring connections can interfere with the operations of vital subsystems.

Parameter drift: Many systems or subsystems can become unserviceable because the components making up the systems have lost integrity. Looseness in even a minor structural joint can allow dangerous relative movement of parts. Resistor or other electronic components can vary in their prime values and so alter the characteristics of the circuit in which they function.

Leakage: Fluid systems depend on the integrity of many joints and seals that can deteriorate or leak. External leakage can cause dangerous loss of fluid content and can risk fire or adverse effects on other components. Hydraulic components can be affected by internal leakage that can affect the power available or speed of operation.

Contamination: Fluid systems are often vulnerable to particulate or other foreign materials causing blockage, deterioration, or other malfunction. Foreign object damage to engines or mechanical interference with control movement within the structure is a form of failure due to contamination.

Software failure: An undetected error, particularly where software has been update or changed can cause system malfunctions that may be critical.

Electromagnetic interference (EMI): Loss of bonding or shielding can allow external electromagnetic fields to interfere with the function of sensitive circuits.

Fraud: Improper documentation can present a part or a process as having integrity that in fact it lacks. Detection of such parts prior to their failure in service requires traceability and recording systems that must have their own integrity.

2.13.2 Causes of Failure

Not all these failure types can be found and remedied by normal maintenance actions at the aircraft although important evidence can often be first detected during routine maintenance. They are however the major concern of the maintenance manager responsible for effective management of the maintenance system. They need to be aware of the root causes of the above failure mechanisms so that their preventative and corrective actions are assured of effectiveness.

The following are some of the more common areas linked to cause of failures:

- Design
- Manufacturing
- Maintenance
- Purchasing
- Operator
- Quality system
- Data
- Sabotage or enemy action

2.13.3 Sources of Failure

It is often not possible to trace back the cause of a failure to its ultimate source. Accident investigations and other enquiries will attempt to trace the cause and effect linkages as far as possible to avoid recurrence of the failure. The following categories illustrate some less obvious potential basic sources of failure that must be of concern to the effective maintenance manager:

- *Ignorance*: A designer, manufacturer, maintainer, or operator is unaware of the risk in the decision being made or the activity being undertaken. Lack of data—a subcategory of ignorance—is one factor over which the maintenance system has some control.
- *Negligence*: While the individual or organization is aware of the corrective action, the measure is not carried through. Errors can be made through inattention. The extreme case is willful negligence, which may be found to be criminal negligence.
- *Poor planning*: Lack of adequate planning can trigger a chain of circumstances leading to a system failure.
- *Sabotage or enemy action*: Where deliberate hostile action is involved.

2.14 MAINTENANCE SYSTEM AND TASKS

As outlined previously, maintenance activities comprise a number of different tasks that can be grouped together as planned maintenance. While there are significant legal, procedural, and terminology differences between civil aviation and military airworthiness systems and procedures, the activities, and much of how they are organized in practice, are very similar. Each system aims at achieving serviceability for the mission, and the signing of a Maintenance Release or Serviceability certificate is required to allow the aircraft to be flown.

2.14.1 Servicing

In a similar way that the private motorcar is put into a local garage for a *servicing*, this term refers to a group of tasks carried out on the compete aircraft. A series of servicing tasks of varying degrees of complexity are normally be performed at predetermined intervals or conditional on some event occurring to the aircraft.

Civil aircraft servicing intervals are approved as a series of "Checks." Friend [27] describes the British Airways scheduled maintenance cycle as "major check, inter check, service check, and ramp check. Airlines may use different terminology. A typical schedule set out by Boeing for the 737 aircraft is shown in Table 2.1.

Some older aircraft have the not-to-exceed flight hours reduced (e.g., the D check is done at 18,000 h). Airlines may seek to combine or split servicing work and some claim additional benefits from tailored programs: for example, in Australia all schedules need CASA approval under CARs 41 and 42.

The military aircraft system for scheduled servicing is very similar to the civil one described earlier. Operational servicing is performed immediately before or after use and may be identified as Before Flight, After Flight, or Turnaround

TABLE 2.1 Typical Service Schedule for Boeing 737

Servicing	Flight hours	Man hours—typical	Details
Transit	Between flights	Low	
A	Daily	Low	Ramp
B	300	12/16	Service
C	1,500	650/700	1–2 elapsed days
Structural	12,000	12,000	Intermediate—items need checking between major intervals
D	24,000	20,000	Major—about every 5 years—takes 25–30 elapsed days

Servicing. This involves checking critical safety items and functions that may have become unserviceable since the last use, checking for damage, replenishment of consumables, and appropriate start-up or shut-down actions. Routine servicing is prescribed in technical maintenance plan (TMP) publications for each type of aircraft at specified intervals and is identified as R1, R2, R4, and so forth in lieu of the civil alphabetical series.

Larger servicing may be carried out periodically, in phases, or under a more flexible but controlled arrangement. The chosen maintenance will depend on the aircraft use.

Routine servicing will be specified for each aircraft type but generally includes the following:

- Functional item checks—specific surveillance and preventative tasks
- Structural inspections of designated items
- Area or zonal inspection
- Functional tests of some items/systems
- Detailed examination of known structural high risk areas
- Sampling of structural areas and installed systems (wiring, piping) of areas of lesser risk
- Surface finish restoration

Special servicing may be required in a series S1, S2, and so forth to meet defined requirements at specific intervals or event points.

2.15 COMPONENT SERVICING

While routine servicing is designated for maintenance of the complete aircraft there is a multitude of components that require removal from the aircraft before they can be worked on.

With more modern systems condition monitoring, built-in-testing or modular design provides for some effective maintenance before removal of the component item. In general, however, components requiring maintenance are sometimes known as MMI's—maintenance managed items, rotable items, or line replaceable units (LRUs). These terms also differentiate items that have to be "managed" from those that are simply replaced when unserviceable, falling into the category of "spares." Component servicing can include overhaul, repair, or bay servicing.

2.16 OVERHAUL

Overhaul of a component, typically an engine or a hydraulic actuator, involves systematic disassembly, inspection, replacement or repair of parts, reassembly, and testing to restore the item to serviceability, generally for a further "overhaul life," if such is specified. The tolerances and standards applied are not necessarily those of a newly manufactured item. Overhaul is usually a preventative maintenance activity applied to items with a defined wear-out pattern, although the process can be used to recover a failed item. With more equipment now designed for on-condition maintenance, the costly process of overhaul is less frequently required. Repair and testing of engines as modules rather than as a complete engine assembly is becoming more common as fundamental items, such as bearings now have longer lives. Lifted components, such as turbine assemblies and compressor disks can be replaced during the servicing of modules rather than requiring a full engine overhaul.

Overhaul facilities require access to extensive supporting workshops, test equipment, and a variety of skills. The process can tie up the high value major component for a long period. Organizations are naturally reluctant to invest in high value spare equipment and the overhaul pipeline has often been the focus of management attention when shortages of serviceable spare engines (say) are likely to leave aircraft on the ground. There is clearly a strong impetus for innovative design and maintenance planning in the overhaul and repair of such items.

2.17 BAY SERVICING

Many items removed from aircraft require a functional test or minor repair to ensure their serviceability. Examples are wheels and brakes that are returned to a bay or workshop for servicing when a tire change or replacement of brake linings is required. A bay servicing is generally a preventative maintenance process to replace parts with a high wear-out rate compared to the rest of the complete item. The maintenance work that can be carried out in such workshops is limited by equipment, facilities, skills available, and the time required. Some items may have a prescribed "bay-servicing" life and are released after servicing.

2.17.1 Repair

Repairs, like all maintenance activities, must be conducted by approved organizations, and by approved personnel using approved processes. This applies particularly to repairs, because by nature they are intended to provide equivalent performance, but they almost invariably change the basic design of the item in some way. The processes must either be specified in the manufacturer's approved repair manual or follow drawings approved by the Regulatory Authority or a delegated design authority.

Repair workshops require special equipment, materials, and experienced tradespeople. As repair events are essentially unscheduled, some repairs may have to be conducted away from home base, requiring a mobile capability, support from other local aircraft organisations or an approved temporary repair scheme to allow the aircraft to return to base satisfactorily. The issue is generally one of safe but temporary operation until a permanent repair can be substituted that will provide for longer-term economical operations. Battle damage repairs are one example of this circumstance that have been extensively researched by military forces.

2.17.2 Modification

Until an aircraft design settles down after development or while it is undergoing changes in role, a number of modifications will be necessary on the aircraft and its fitted equipment items. The fitting of minor modifications is done usually during a period of scheduled maintenance downtime. The modification processes must be controlled for quality and approved in the same way as repairs. Some modifications will have a greater urgency than others and the period of time before they must be fitted can depend on a range of factors, including safety, reliability, compatibility, and operational economics. The regulatory authority may insist on an urgent change and issue a Mandatory Airworthiness Directive to authorize this modification. Special servicing may also be directed in this way.

In the case of major modifications or safety inspections requiring extensive maintenance work, a special project may be undertaken to apply the processes of project management to coordination of all aspects of this task; in this way planning and direction of various operational, supply, logistic, facilities, and personnel can be closely managed to achieve optimum results with minimum disruption to operations.

2.18 REPLACEMENT/THROWAWAY

Many aircraft items that have failed, worn out, or deteriorated are economically unrepairable. In this case the item is discarded to scrap and replaced by a serviceable item obtained as a spare part. The criteria for a decision on economic reparability may be affected in the short term by the lack of availability of the replacement part. As part of the Logistic Support Analysis process, decisions are usually made early in the planning for aircraft maintenance that such items will not be repaired, and a source of supply of spares will be provisioned to meet such replacement needs within an acceptable lead time. The quality of replacements must be assured with the same concern as for the basic aircraft and particular care taken to avoid problems with "bogus parts" that may be supplied without the normal quality controls needed to ensure integrity of the part.

2.19 FUNCTIONAL TESTING

Operational checks and functional testing are surveillance maintenance tasks. The former may be a simple check during a servicing to ensure an item is operating. A functional test is a more thorough task in which operating parameters are measured against a specified standard. For example, the complex adaptive flight control system of the F-111 is tested in the servicing hangar using an electrohydraulic tester that conducts a series of tests that assure the correct functioning of all system components. The built-in test equipment of modern avionic systems can perform similar checks in flight and regularly transmit the serviceability status to crew and ground stations.

Functional flight tests have been a feature of maintenance on some aircraft types where ground testing is unable to represent the conditions adequately or an additional assurance of correct operation is needed.

2.20 CALIBRATION

This important surveillance maintenance function is used to detect unacceptable changes in, or provide regular assurance of, the performance of an item. Recalibration of the compass in an aircraft as an example is critical to navigation; magnetic field changes in the aircraft can affect the performance of this sensor. Calibration of precision measuring equipment used on aircraft and components is often critical to maintenance process integrity. A separation is maintained between the calibration tasks and of any subsequent readjustment or repair recorded as an additional maintenance action.

2.21 NONDESTRUCTIVE EVALUATION

This is a surveillance maintenance process that enables item integrity to be checked without affecting item serviceability. Since the outcome of an nondestructive evaluation (NDE) check is often critical to airworthiness, there are often special training and qualification requirements laid down. This topic will be covered further in later sessions.

2.22 AVIONICS MAINTENANCE

Increasing use of electronics and digital computer technology in the primary systems of an aircraft has extended the scope of maintenance and related engineering tasks. No longer it is possible to assign an "accessory" status to the avionics systems in an aircraft that contribute directly to flight safety in such areas as flight control and engine fuel control. With increasing needs of communication and navigation systems in safe airways control, failures of avionics systems have become critical concerns of the maintenance system.

It is especially in this area, where failures tend to be random, that design for redundancy and reliability-based maintenance practices have gained urgency. Computer-based automatic test equipment and built-in test facilities have tended to offset the problems of surveillance inspection of such systems. These will be discussed in later sections.

2.23 SOFTWARE MAINTENANCE

The term *software maintenance* is something of a misnomer because software maintenance is essentially a continuation of the developmental design process. Failures in airborne software are normally found in unusual operational modes—"bugs" that need to be redesigned out of the system, having been undetected in initial development testing. There remains a significant continuing airworthiness and thus safety concerns, however, that airborne software management processes sustain the same level of integrity as all other critical aircraft systems. It is unusual for civil operators to have authority to modify their own software while some military systems have been established on the basis of experience dating back to the 1960s.

One software characteristic is that a small change introduced in one part of the program, perhaps late in development or aimed at resolving a problem, may have serious implications in other, apparently unrelated areas of code. Operational improvements may be sought by way of similar, apparently minor changes.

Detailed consideration of this topic is beyond the scope of this course. It is important, however, to appreciate the need for careful risk management in tackling airborne software problems and also the need for observing the same critical standards for making and recording changes to the software as applied in the original design, development, and testing.

2.24 INTERDEPENDENCE OF OPERATIONS AND MAINTENANCE

There are many considerations affecting the planning for maintenance activities of a fleet of aircraft. The maintenance manager must take these into account, as well as engineering and airworthiness considerations.

2.24.1 Factors Affecting the Airline's Maintenance System

There are many factors to be taken into account in devising a maintenance system for an airline, as discussed in the following sections.

2.24.1.1 Seasonal Traffic Trends

As many aircraft as possible should be available for service during periods of high traffic demand, while aircraft can be taken out of service for maintenance during periods of low traffic. High traffic periods are often of short duration, such as Easter and Christmas, but there can be a marked variation between summer and winter traffic—in Europe, for example, the ratio has reached almost two to one. All these factors must be taken into account to assess the availability of aircraft for maintenance; difficulty arises in equalizing the workload for maintenance bases.

2.24.1.2 Geography of the Operation

Is the pattern of aircraft operation radial? For example, do most, if not all, aircraft return to the same base each night? This would tend to happen in a European operation based in London.

Is aircraft operation a "through' operation? For example, does it comprise a main trunk line with loops and feeders coming into it at various points, perhaps skipping the odd point sometimes on the main trunk route. This is more often the pattern of Australian domestic airlines.

Is the operation long haul, with aircraft away from the main base for days at a time? Qantas airline is an example of a long-haul operation. Some sectors will be more heavily loaded than others.

What about curfews? How might they limit operations?

Is it necessary to station aircraft overnight at ports away from the main base? If so, what maintenance facilities, if any, do such ports offer?

2.24.1.3 Location and Size of Maintenance Establishments

Determine the location and size of facilities, staff, and capabilities available at each port.

2.24.1.4 Size and Composition of the Airline Fleet

The numbers of each type of aircraft must be known. With larger fleets it is possible to spread the load more evenly but nevertheless total load must be assessed against division establishments. It may also be necessary to set up a different system for one aircraft type. If the fleet is small, the consequences of one aircraft out of service for maintenance are far more serious than for a large fleet.

2.24.1.5 Aircraft Utilization

This will have an important bearing on the time available for maintenance and consequently the work load imposed on the division; turn-around times may be, such as to require more staff at each port.

2.24.1.6 Weather

Apart from having an important effect on utilization and delays, plans must be made to provide flexibility in the event of serious dislocation of services. Again

North America and European winters seriously curtail flying activities and indeed deter passengers from traveling by air if such weather causes the services to be unreliable. For example, weather causes Australian Domestics much less trouble than other airlines, but Qantas does have problems in its services to Europe, in particular.

2.24.1.7 Availability of Subcontracting for Servicing and Maintenance

Considerable work within the maintenance division is subcontracted out, but a large proportion of this would not affect the maintenance system used, but would more likely affect the inventory of spare parts, depending on subcontractor turnround times and reliability. However, it is possible some subcontracting could directly affect airline operations. There are two types of subcontractor in this field:

1. The general aircraft engineering type of company.
2. Another airline operator who may or may not be a competitor.

The sort of work that, if subcontracted, could endanger on-time performance would include: major block maintenance, engine overhaul, undercarriage overhaul, and the like. There must be significant economic justification for putting such work out, as once it is sent out control of "on-time" performance is considerably weakened. Assessment must be made of the subcontractor's past performance and their industrial relations. If they are a direct competitor, who would receive priority? Airline operations are so heavily dependent on skilled labor that industrial trouble can be disastrous. In these situations extra spare parts would need to be kept in the inventory, and monitoring of subcontractors would be essential. For these reasons, all larger airlines try to carry out all major work themselves, to retain control of the operation; the cost to an airline of having an aircraft unexpectedly on the ground can far outweigh the cost differences between the two approaches.

2.24.1.8 Competitors' Operations

Competitors' operations will be watched by the whole airline and every effort must be made to avoid giving traffic to a competitor due to lack of serviceable aircraft, not only because aircraft is not available, but also caused by malfunction leading to unserviceability.

2.24.1.9 Availability of Staff

Suitably trained staff must be available at all ports where maintenance work is to be performed. They must be in sufficient numbers and the organization must be flexible enough to cope with any emergencies. The main ports must be ready to fly staff to ports without personnel if a fault develops that must be repaired before the next flight: not all the fail–safe principles in the world can ensure that this will not occur. Airlines can only minimize these kinds of events.

2.24.2 Factors Affecting the Military Maintenance System

In military operations, very similar considerations apply. Factors that affect planning include:

- What is the annual rate of operational effort? Expressed in flying hours for the fleet per annum, this rate sets budgetary allowances for all resources, including contract maintenance, spares, and personnel. This rate also determines the notional ceiling for the average weekly flying rate for available aircraft. Factors beyond the operator's control, such as crew shortages, that may impede the attainment of this objective must be taken into account during planning.
- Are there planned periods of more intensive operational activity, such as exercise deployments that could be affected by aircraft or personnel availability? Efforts needed for a limited period can be enhanced by forward planning, to increase availability beyond the normal average level.
- What is the type of operation? Are fighter aircraft required to operate in pairs or larger formations? Are transport aircraft scheduled for long local missions or overseas tasks? What special equipment, fuel tanks, or weapons loads are required?
- Are all aircraft equally available, or must some be managed to fit external or internal planned modification or maintenance programs?

The tools available to assist in maintenance planning and control to meet operational mission requirements will be the subject of later chapters in this book.

Chapter 3

Aircraft Reliability and Maintainability Analysis and Design

Chapter Outline

Reliability Based Aircraft Maintenance Optimization and Applications
http://dx.doi.org/10.1016/B978-0-12-812668-4.00003-4

3.1 RELIABILITY FUNDAMENTAL MATHEMATICS

This chapter gives an overview of the important stochastic failure distributions and reliability analysis models.

3.1.1 Density Function

Suppose we observe the system at times t_1, t_2, and so forth, and we then define the failure density function as follows:

$$f(t) = \frac{N_s(t_i) - N_s(t_{i+1})}{N_0 \cdot (t_{i+1} - t_i)}, \quad t_i < t \le t_{i+1} \tag{3.1}$$

This is the ratio of the number of failures occurring in the interval to the size of the original population, divided by the length of the interval.

3.1.2 Failure Probability Function

Consider the lifetime, t, of a unit of a certain type from its installation until it fails, that is, to the point when it is unable to perform its intended function. We put $t = 0$ initially. It is natural to assume that the lifetime T is a stochastic variable because we cannot say in advance how long it will function.

Let $F(t)$ denote the lifetime distribution of t, that is: $F(t) = P(T \le t)$. In the following we assume T has a continuous distribution, such that:

$$F(t) = \int_0^t f(t)\,dt \tag{3.2}$$

The function $f(t)$ is called the probability density of T. The relation between $F(t)$ and $f(t)$ is illustrated in Fig. 3.1.

In view of the frequency interpretation of the probability concept, we interpret $F(t)$ as the portion of units that will fail within t units of time, when a large number of units of this type is installed at a time $t = 0$.

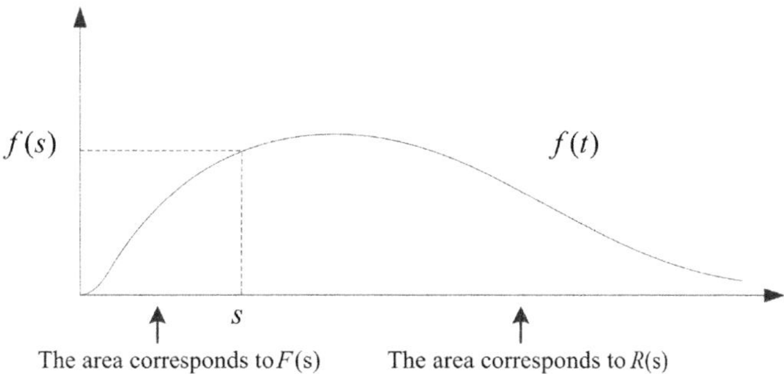

FIGURE 3.1 **Relation between $F(t)$ and $f(t)$.**

3.1.3 Failure Rate

Similarly, failure rate could be defined as: a ratio of the number of failures occurring in the time interval to the number of survivors at the beginning of the time interval divided by the length of the time interval.

$$\lambda(t)=\frac{N_s(t_i)-N_s(t_{i+1})}{N_s(t_i)\cdot(t_{i+1}-t_i)},\quad t_i<t\leq t_{i+1} \tag{3.3}$$

Failure rate has the following relationship with density and reliability functions.

$$\lambda(t)=\frac{f(t)}{R(t)} \tag{3.4}$$

3.1.4 Reliability Function

Often we are more interested in $P(T > t)$, we therefore introduce a special notation for this probability:

$$R(t)=P(T>t)=1-F(t) \tag{3.5}$$

We call $R(t)$, the reliability (survivor) function.

3.1.5 Bathtub Curve

A typical shape of the failure rate is the so-called bathtub curve, which is shown in Fig. 3.2.

A plot of instantaneous failure rate versus time is known as a hazard curve. It is more often called a bathtub curve due to its shape.

1. The curve was first used in the life insurance industry.
2. It is the result of plotting the human death rate over time.
3. Engineers apply this concept to repairable systems as well as nonrepairable components and one-shot devices.
4. The horizontal axis shows age of the product.

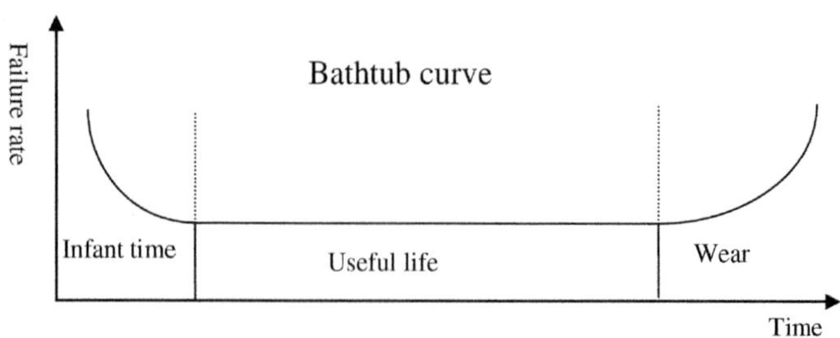

FIGURE 3.2 **Bathtub curve.**

5. It represents time or life, t.
6. The vertical axis depicts the hazard rate or instantaneous failure rate.
7. The bathtub curve illustrates three phases of a product's life:
   a. Infant mortality is the front portion of the curve.
   b. Random/chance failure is the mid portion of the curve.
   c. Wear-out time is the tail portion of the curve.

The infant mortality region of the curve represents parts failing early in the product life-span.

The infant mortality portion of the failure rate curve takes its shape from the *early life failure curve*, which represents a *decreasing* failure rate over time. This is caused by improper manufacturing, assembly, and material problems.

The middle part of the curve or *useful life region*, represents parts failing randomly.

The useful life portion of the failure rate curve takes its shape from the *random stress-related failures line*, which represents a *constant* failure rate over time. This is caused by inadequate design strength given the stress encountered.

3.1.6 MTTF

The mean time to failure (MTTF) is defined by:

$$ET = \int_0^\infty tf(t)\,dt = \int_0^\infty R(t)\,dt \tag{3.6}$$

Thus, the mean lifetime is given by area below the survivor function $R(t)$. The quantity ET is often referred to as the mean time to failure (MTTF).

3.2 SOME COMMON FAILURE DISTRIBUTIONS

In this section, a few often used and common distributions are introduced.

3.2.1 Exponential Distribution

If the lifetime T is said to be exponentially distributed with parameter $\lambda(>0)$, then the density function follows:

$$f(t) = \lambda e^{-\lambda t}, \quad or \quad f(t) = \frac{1}{\theta} e^{-\frac{t}{\theta}}, \quad 0 < \lambda < \infty, 0 \le t < \infty \tag{3.7}$$

$$F(t) = P(T \le t) = 1 - e^{-\lambda t}, \quad t \ge 0 \tag{3.8}$$

Thus the exponential distribution is characterized by a constant failure rate (λ). A unit having an exponential failure time distribution has a tendency to failure that does not depend on the unit age. Assuming that the unit has survived u hours, the probability that the unit then will survive additional v hours is given by:

$$
\begin{aligned}
P(T > u+v \mid T > u) &= \frac{P(T > u+v \cap T > u)}{P(T > u)} \\
&= \frac{P(T > u+v)}{P(T > u)} \\
&= \frac{e^{-\lambda(u+v)}}{e^{-\lambda u}} \\
&= e^{-\lambda v} \\
&= P(T > v)
\end{aligned}
\tag{3.9}
$$

Thus the probability of survival of the additional v hours is not dependent on how long the unit has functioned. The exponential distribution is the only distribution with this property. The lack of memory simplifies the mathematical modelling.

The fact that the failure rate is constant for large values of t may seem unrealistic for practical applications. We must, however, remember we are usually interested in studying the lifetime in a limited time period. The failure rate assumed outside this period will then not be critical. In addition, the probability that the unit will really last so long will be small, since the mass of the probability density function is positioned around a "small" value of t (Fig. 3.3).

The exponential distribution usually gives a good description of the lifetime of electrical and electronic units. In some cases, it has also been useful for modelling units comprising a large number of mechanical components—for example, pumps—where the unit has been in operation for a relatively long period of time and maintenance has led to different ages of various components of the unit.

The mean and variance in the exponential distribution are given by:

$$ET = \text{MTBF} = \frac{1}{\lambda} \tag{3.10}$$

Thus the exponential represents a lower limit for the survivor function $R(t)$ when $t < ET$. Assuming an exponential distribution when the true distribution is another IFR distribution, gives an underestimated value of $R(t)$ for these values of t.

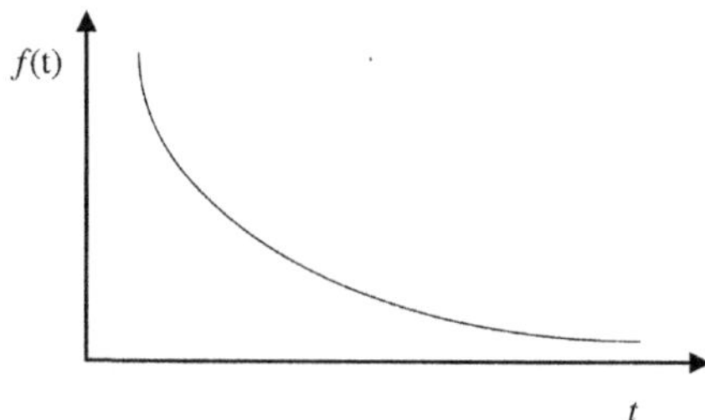

FIGURE 3.3 Density function of the exponential distribution.

3.2.2 Weibull Distribution

The density function of Weibull distribution is as follows:

$$f(t)=\frac{\beta}{\eta}\left(\frac{t-\gamma}{\eta}\right)^{\beta-1}\exp\left[-\left(\frac{t-\gamma}{\eta}\right)^{\beta}\right] \tag{3.11}$$

The Weibull failure distribution function is:

$$\begin{aligned} F(t)&=\int_{\gamma}^{t}\frac{\beta}{\eta}\left(\frac{t-\gamma}{\eta}\right)^{\beta-1}\exp-\left[\left(\frac{t-\gamma}{\eta}\right)^{\beta}\right]\cdot dt \\ &=1-e^{-\left(\frac{t-\gamma}{\eta}\right)^{\beta}} \end{aligned} \tag{3.12}$$

The reliability function would be:

$$\begin{aligned} R(t)&=1-F(t) \\ &=e^{-\left(\frac{t-\gamma}{\eta}\right)^{\beta}} \end{aligned} \tag{3.13}$$

Where the β, η, and γ are the shape, scale, and position parameters, respectively, and the $\Gamma()$ is the Gamma function. In practice, the position parameter γ is very small. Sometimes we consider $\gamma = 0$. Then the failure distribution function could be:

$$F(t)=1-e^{-\left(\frac{t}{\eta}\right)^{\beta}}=1-e^{-\lambda t^{\beta}} \tag{3.14}$$

If we choose $\beta = 1$, then the failure rate becomes a constant. Hence the exponential distribution is a special case of the Weibull distribution. In the following figure the density function and failure rate are shown graphically for some values of the parameters (Fig. 3.4).

Let $t = \eta$, then the reliability $R(t) = e^{-1} = 0.3679$ and the failure probability is $F = 1 - R = 0.6321$. The quantity of $t = \eta$ is often called the characteristic lifetime.

3.2.3 Normal Distribution

The normal distribution is probably the most important and widely used distribution in the entire field of statistics and probability. Although it has some important applications in reliability evaluation, it is less significant in this field than many other distributions. The normal distribution is sometimes used to model wear-out failure, but it is also used as a lifetime distribution for batteries and condensers.

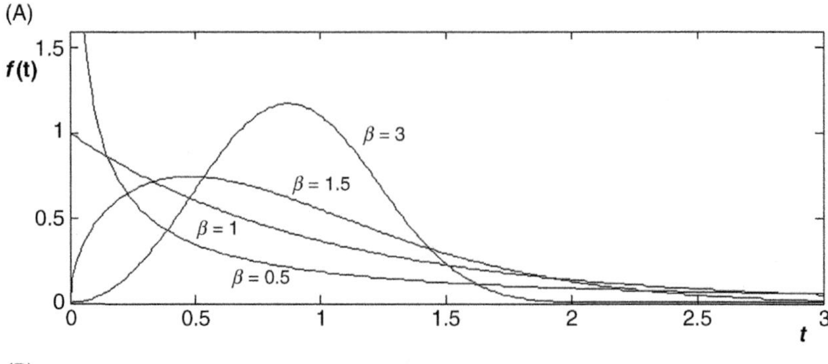

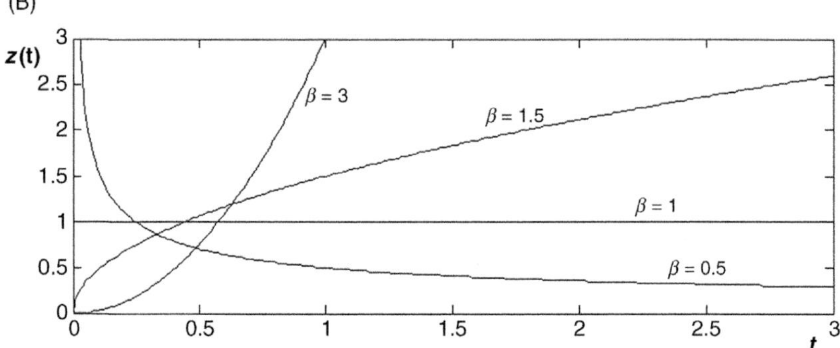

FIGURE 3.4 (A) Density function and (B) failure rate of the Weibull distribution.

The density function of normal distribution

$$f(t) = \frac{1}{\sigma\sqrt{2\pi}}\exp\left[-\frac{(t-\mu)^2}{2\sigma}\right], \quad -\infty < t < +\infty \tag{3.15}$$

and the failure distribution function

$$\begin{aligned} F(t) &= P(T \le t) \\ &= \frac{1}{\sigma\sqrt{2\pi}}\int_{-\infty}^{t}\exp\left[-\frac{(t-\mu)^2}{2\sigma}\right]dt \end{aligned} \tag{3.16}$$

Let $z = t - u/\sigma$, $\mu = 0$, $\sigma = 1$, then we can get the standard normal distribution:

$$\begin{aligned} \Phi(t) &= \frac{1}{\sqrt{2\pi}}e^{-\frac{t^2}{2}} \\ \Phi(t) &= \frac{1}{\sqrt{2\pi}}\int_{-\infty}^{t}e^{-\frac{z^2}{2}}dz \end{aligned} \tag{3.17}$$

3.2.4 Lognormal Distribution

The density function of the lognormal distribution is:

$$f(t) = \frac{1}{\sigma_1\sqrt{2\pi}}\exp\left[-\frac{(\ln t-\mu_1)^2}{2\sigma_1}\right], \quad -\infty < t < +\infty \tag{3.18}$$

The distribution function follows as:

$$\begin{aligned} F(t) &= P(T \le t) \\ &= \frac{1}{\sigma_l\sqrt{2\pi}}\int_{-\infty}^{t}\exp\left[-\frac{(\ln t-\mu_1)^2}{2\sigma_1}\right]dt \\ &= \Phi\left(\frac{\ln t-\mu_1}{\sigma_1}\right) \end{aligned} \tag{3.19}$$

Then the mean and variance of t variable are μ and σ, respectively:

$$\begin{aligned} \mu &= E(t) = \exp\left[\mu_1 + \frac{1}{2}\sigma_1^2\right] \\ \sigma^2 &= Var(t) = \mu^2\left(e^{\sigma_1^2}-1\right) \end{aligned} \tag{3.20}$$

3.2.5 Summary of Often Used Distributions

The summary of often used distributions is given in Table 3.1.

3.3 BINARY SYSTEM RELIABILITY MODELS

Assume the binary variable:

$$A = \begin{cases} 1, & \textit{if the system is in the functioning state} \\ 0, & \textit{if the system is in the failure state} \end{cases}$$

3.3.1 Series System

Reliability block diagram and reliability model:

If R_1, R_2, and R_n, respectively, are the reliability of the components in a series system, then the system reliability could be calculated as (Fig. 3.5):

$$R_s = P(R_1 \cdot\cap R_2 \cdot\cap\cdots\cap R_n) = R_1 \cdot R_2 \cdot\cdots\cdot R_n$$

$$R_s = \prod_{i=1}^{n} R_i \tag{3.21}$$

TABLE 3.1 Summary of Often Used Distributions

Name	Parameters	Density function	Mean	Variation
Normal	$-\infty < \mu < \infty$ $\sigma > 0$	$f(t) = \frac{1}{\sigma\sqrt{2\pi}} \exp\left[-\frac{(t-\mu)^2}{2\sigma}\right], \quad -\infty < t < +\infty$ $F(t) = \frac{1}{\sigma\sqrt{2\pi}} \int_{-\infty}^{t} \exp\left[-\frac{(t-\mu)^2}{2\sigma}\right] dt$	μ	σ^2
Log Normal	$-\infty < \mu_l < \infty$ $\sigma_l > 0$	$f(t) = \frac{1}{\sigma_l\sqrt{2\pi}} \exp\left[-\frac{(\ln t-\mu_l)^2}{2\sigma_l}\right], \quad -\infty < t < +\infty$ $F(t) = \frac{1}{\sigma_l\sqrt{2\pi}} \int_{-\infty}^{t} \exp\left[-\frac{(\ln t-\mu_l)^2}{2\sigma_l}\right] dt$ $= \Phi\left(\frac{\ln t - \mu_l}{\sigma_l}\right)$	$\exp\left[\mu_l + \frac{1}{2}\sigma_l^2\right]$	$\mu^2\left(e^{\sigma_l^2} - 1\right)$
Weibull	$\beta > 0$ $\gamma \geq 0$ $\eta > 0$	$f(t) = \frac{\beta}{\eta}\left(\frac{t-\gamma}{\eta}\right)^{\beta-1} \exp\left[-\left(\frac{t-\gamma}{\eta}\right)^{\beta}\right]$ $F(t) = \int_{\gamma}^{t} \frac{\beta}{\eta}\left(\frac{t-\gamma}{\eta}\right)^{\beta-1} \exp\left[-\left(\frac{t-\gamma}{\eta}\right)^{\beta}\right] \cdot dt$ $= 1 - e^{-\left(\frac{t-\gamma}{\eta}\right)^{\beta}}$	$\gamma + \eta\Gamma\left(\frac{1}{\beta} + 1\right)$	$\eta^2\left\{\Gamma\left(\frac{2}{\beta}+1\right) - \left[\Gamma\left(\frac{1}{\beta}+1\right)\right]^2\right\}$

Exponential	$\lambda > 0$	$f(t) = \lambda e^{-\lambda t}$, *or* $f(t) = \frac{1}{\theta} e^{-\frac{t}{\theta}}, 0 < \lambda < \infty, \quad 0 \le t < \infty$ $F(t) = 1 - e^{-\lambda t}, \quad t \ge 0$	$\frac{1}{\lambda}$ θ	$\frac{1}{\lambda^2}$ θ^2
Gamma	$n > 0$ $\lambda > 0$	$f(t) = \frac{\lambda}{\Gamma(n)} (\lambda t)^{n-1} e^{-\lambda t}, \quad t \ge 0$ $F(t) = 1 - R(t) = 1 - \sum_{i=0}^{n-1} \frac{(\lambda t)^i}{i!} e^{-\lambda t}$	$\frac{n}{\lambda}$	$\frac{n}{\lambda^2}$
Minimum Extreme	$-\infty < \mu < \infty$ $\sigma > 0$	$f(t) = \frac{1}{\sigma} e^{\frac{t-\mu}{\sigma}} e^{-e^{\frac{t-\mu}{\sigma}}}$ $F(t) = 1 - e^{-e^{\frac{t-\mu}{\sigma}}}$	$\cong \mu - 0.5772\sigma$	$\frac{\pi^2}{6}\sigma^2$
Maximum Extreme	$-\infty < \mu < \infty$ $\sigma > 0$	$f(t) = \frac{1}{\sigma} e^{-\frac{t-\mu}{\sigma}} e^{-e^{-\frac{t-\mu}{\sigma}}}$ $F(t) = e^{-e^{-\frac{t-\mu}{\sigma}}}$	$\cong \mu + 0.5772\sigma$	$\frac{\pi^2}{6}\sigma^2$

FIGURE 3.5 Series system.

If the components distribute as exponential distribution, $\lambda = 1/MTBF$, then the reliability of the component would be:

$$R(t) = e^{-\lambda t} = e^{-\frac{t}{\text{MTBF}}} \tag{3.22}$$

Then the system reliability is denoted as:

$$R_s(t) = \prod_{i=1}^{n} e^{-\lambda_i t} = e^{-\left(\sum_{i=1}^{n} \lambda_i\right)t} \tag{3.23}$$

The system failure rate would be:

$$\lambda_s = \sum_{i=1}^{n} \lambda_i \tag{3.24}$$

Example: -*Question*

An electronic circuit consists of 5 silicon transistors, 10 silicon diodes, 20 composition resisters, and 5 ceramic capacities in continuous series operation, and assumes that under the actual stress conditions in the circuit the components have the following failure rates:

- Silicon transistors: $\lambda_t = 0.000008\ \text{h}^{-1}$
- Silicon diodes: $\lambda_d = 0.000002\ \text{h}^{-1}$
- Composition resistors: $\lambda_r = 0.000001\ \text{h}^{-1}$
- Ceramic capacitors transistors: $\lambda_c = 0.000004\ \text{h}^{-1}$

Estimate the reliability of this circuit for 10 h of operation.

Solution:

- Circuit system failure rate is: $\lambda_s = 5\lambda_t + 10\lambda_d + 20\lambda_r + 5\lambda_c = 0.0001\ \text{h}^{-1}$
- The reliability at 10 h is $R(10) = \exp(-0.0001 \cdot 10) = 0.999 = 99.9\%$

3.3.2 Parallel System

The parallel system model is (Fig. 3.6):

$$R_s = P(R_1 \cdot U\, R_2 \cdot U \cdots U\, R_n) = 1 - (1 - R_1) \cdot (1 - R_2) \cdot \cdots \cdot (1 - R_n) \tag{3.25}$$

If the components in the system are independent and distributed as exponential, then:

$$R_s(t) = 1 - \prod_{i=1}^{n} \left(1 - e^{-\lambda_i t}\right) \tag{3.26}$$

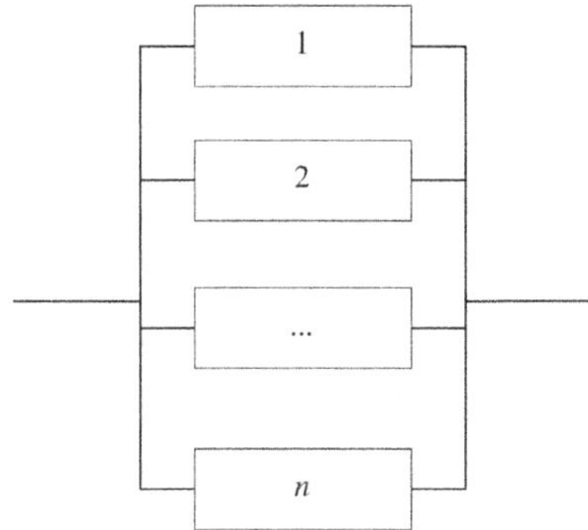

FIGURE 3.6 **Parallel system.**

If the failure rate λ is constant, then system reliability could be denoted as:

$$R_s(t) = 1 - \left(1 - e^{-\lambda t}\right)^n \tag{3.27}$$

3.3.3 Standby Redundancy System

Standby redundancy involves additional units that are activated only when the operating unit fails.

Case 1:

Some assumptions (Fig. 3.7):

- The means of sensing that a failure has occurred and for switching from the defective to the standby unit is assumed to be failure free.
- The standby units are assumed to have identical, constant failure rates to the main unit.
- The standby units are assumed not to fail while in the idle state

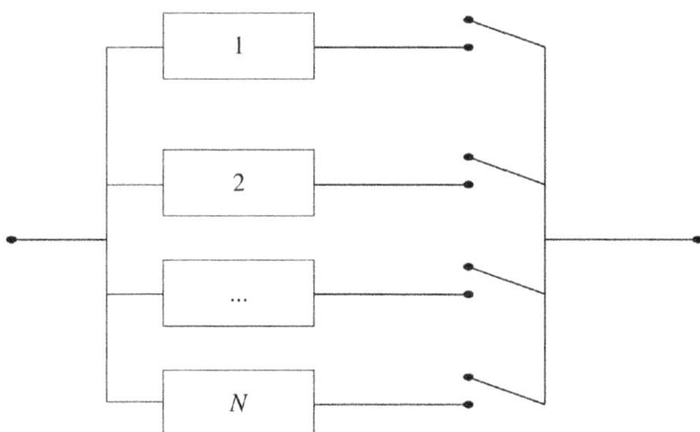

FIGURE 3.7 **Standby redundancy system.**

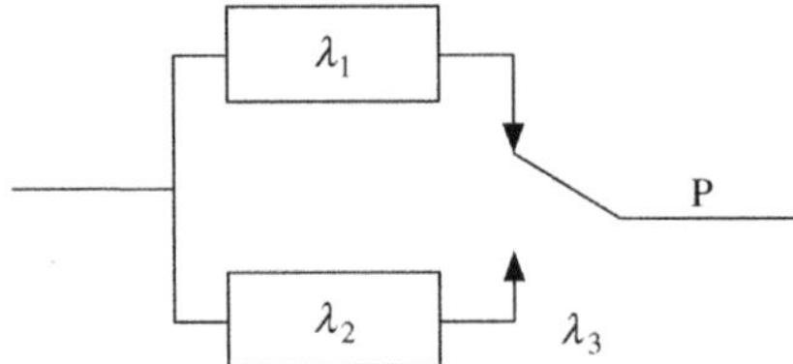

FIGURE 3.8 Two units model.

- The detective units are assumed perfect. No repair is initiated until the system has failed.
- The reliability is then given by the first n terms of the Poisson expression:

$$R_{\text{system}}(t) = e^{-\lambda t}\left(1+\lambda t+\frac{\lambda^2 t^2}{2!}\cdots\frac{\lambda^{(n-1)}t^{(n-1)}}{(n-1)!}\right) \tag{3.28}$$

If just for units, this reduces to:

$$R_{\text{system}}(t) = e^{-\lambda t}(1+\lambda t) \tag{3.29}$$

Case 2: General Model

Fig. 3.8 shows a general case model of two units with some of the above assumptions removed. In the figure:

- λ_1 is the constant failure rate of the main unit
- λ_2 is the constant failure rate of the standby unit when in use
- λ_3 is the constant failure rate of the standby unit in the idle state
- P is the one-shot probability of the switch performing when required.

The reliability is given by:

$$R_{\text{system}}(t) = e^{-\lambda_1 t} + \frac{P\lambda_1}{\lambda_2-\lambda_1-\lambda_3}\left(e^{-(\lambda_1+\lambda_3)t} - e^{-\lambda_2 t}\right) \tag{3.30}$$

3.4 MECHANICAL RELIABILITY—STRESS–STRENGTH INTERFERENCE MODEL

3.4.1 Introduction of Theory

The *stress–strength interference method* gives the definition of structure failure as the imposed stress (load) exceeds the strength (capability) of structure. Failure probability or unreliability is the probability that the stress is greater than the strength. The stress–strength interference method may be used in conjunction with a variety of failure modes, such as yielding, buckling, fracture, and fatigue.

Fig. 3.9 shows the theory of the stress–strength interference method, that is, the probability density functions of stress and strength and their interference

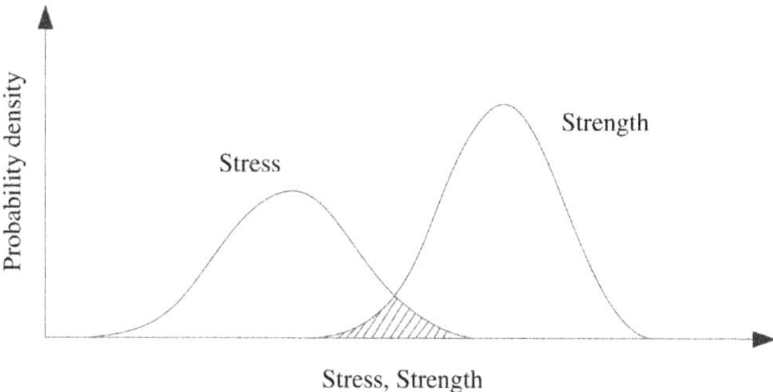

FIGURE 3.9 **Graphical representation of stress–strength interference.**

(overlap). It should be pointed out that the overlapped area (shown shaded) is not equal to the failure probability. However, this area is qualitatively proportional to the failure probability (the larger the area, the higher the failure probability) as long as the mean value of stress is less than the mean value of strength. The failure probability is equal to the black area in Fig. 3.9.

3.4.2 Analytical Results

If we let C and L represent the Capability (strength) and Load (stress) respectively, then variable U can be defined by:

$$U = C - L \tag{3.31}$$

The safety margin equation is:

$$U = C - L \begin{cases} when\, U > 0, & structure\, safety \\ when\, U \leq 0, & structure\, failure \end{cases} \tag{3.32}$$

The failure probability of a structure is given by:

$$P_t = P[U < 0] = \int_{-\infty}^{0} f_U(u)\,du = F_U(0) \tag{3.33}$$

Where $f_U(u)$ is the density function of variable U.

Normally, the stress x and strength y satisfy the normal distribution.

$$P(x) = \int \frac{1}{\sigma_x\sqrt{2\pi}} \cdot \exp\left[-\frac{(x-\mu_x)^2}{2\sigma_x^2}\right] \cdot dx \tag{3.34}$$

$$P(y) = \int \frac{1}{\sigma_y\sqrt{2\pi}} \cdot \exp\left[-\frac{(y-\mu_y)^2}{2\sigma_y^2}\right] \cdot dy \tag{3.35}$$

If the x and y are independent random variables, the $U = y - x$ is a random variable as well and also satisfies the normal distribution.

$$P(U) = \int \frac{1}{\sigma_U \sqrt{2\pi}} \cdot \exp\left[-\frac{(y-\mu_U)^2}{2\sigma_U^2} \right] \cdot dU \tag{3.36}$$

The mean and variation are respectively following:

$$\begin{aligned} \mu_U &= \mu_y - \mu_x \\ \sigma_U &= \sqrt{\sigma_x^2 + \sigma_y^2} \\ Z &= \frac{\mu_U}{\sigma_U} = \frac{\mu_y - \mu_x}{\sqrt{\sigma_x^2 + \sigma_y^2}} \end{aligned} \tag{3.37}$$

Reliability would be:

$$R = 1 - P = 1 - \Phi(-Z) = \Phi(Z) \tag{3.38}$$

3.4.3 Example

Question: The stress L and strength C of a steel rope all satisfy normal distribution. L(544300, 113400), C(907200, 90700), calculate the reliability of the steel rope.

Solution:
Reliability index:

$$\begin{aligned} \beta &= \frac{544300 - 907200}{\sqrt{13400^2 + 90700^2}} = -\frac{362900}{916845.51} \\ &= -3.958 \end{aligned}$$

Reliability is:

$$R = \Phi(-\beta) = 0.99999$$

3.5 FUZZY RELIABILITY THEORY

3.5.1 Irrationality of Conventional Reliability Theory

Binary Assumption

The fault or failure of a device is in only two situations, which are either "complete normal" or "complete failure." The failure function can be denoted by (0,1).

Probability assumption

1. The events should be described clearly.
2. There are a large number of samples in the experiment.
3. The outcomes must be repeatable from experiment to experiment. The outcome of one trial does not influence the outcome of a previous or future trial.
4. There is no influence from humans in the trial. All the trials should be independent.

In solving some engineering problems, these four presuppositions cannot be satisfied at the same time.

3.5.2 Fuzzy Reliability Basic Theories

A fuzzy fault can be defined as a product loss of desired function at some level of degree. As to the no-repaired products, fuzzy fault could be called fuzzy failure. The biggest difference between a fuzzy fault and a conventional fault is that a fuzzy fault would describe the degree of the fault, but the conventional fault would not.

- Possibility assumption
- Fuzzy set and membership

$$\tilde{A}=\left\{\left(\omega,\mu_{\tilde{A}}(\omega)\right):\omega\in\Omega;\mu_{\tilde{A}}(\omega)\in[0,1]\right\}$$

where μ is the degree of membership of ω in the set.

3.5.3 Fuzzy Reliability

Assume membership function, for example, trapezoid distribution:

$$\mu_{\tilde{A}}(x)=\begin{cases} 1, & (0\le x\le a_1) \\ \dfrac{a_2-x}{a_2-a_1}, & (a_1\le x\le a_2) \\ 0, & (a_2<x) \end{cases}$$

Fuzzy reliability:

$$\tilde{R}(\tilde{A})=\mu_{\tilde{A}}\cdot R$$

3.5.4 Fuzzy Failure Rate

$$\begin{aligned}\tilde{\lambda}(t)&=\lim_{\Delta t\to 0}\frac{P\left(\tilde{B}_2\mid\tilde{B}_1\right)}{\Delta t}\\ &=\lim_{\Delta t\to 0}\frac{\tilde{F}(t+\Delta t)-\tilde{F}(t)}{\Delta t}\cdot\frac{1}{\tilde{R}}\\ &=\frac{\tilde{F}(t)}{\tilde{R}}\end{aligned}$$

3.5.5 Fuzzy MTBF

$$\begin{aligned}\overleftrightarrow{\text{MTTF}} &= \int_0^\infty t\cdot\left(\frac{d\tilde{F}(t)}{dt}\right)\cdot dt \\ &= \int_0^\infty t\cdot\left(\frac{d\left(1-\tilde{R}(t)\right)}{dt}\right)\cdot dt \\ &= \int_0^\infty t\cdot d\tilde{R} \\ &= \int_0^\infty \tilde{R}(t)\cdot dt\end{aligned}$$

3.6 HARDWARE RELIABILITY

Hardware reliability is often defined as the probability that the equipment will perform throughout the intended mission life, within specified tolerance under specified life cycle loads. Reliability is not a matter of chance. It has to be consciously and actively built into hardware through careful specification of good design and manufacturing practice. The reliability assessment during the design phase includes:

- Reliability allocation: based on complexity, cost, and risk
- Feasibility evaluation
- Determination of deficiencies in current database regarding material properties, application profile, and field failure data
- Comparison of alternative design configuration, based on relative reliability margins
- Comparison of alternative manufacturing processes, based on relative reliability margin
- Evaluation of cost effectiveness, based on the reliability margin
- Development of tradeoffs with other product parameters, such as cost, risk
- Development of time, producibility, and maintainability
- Design of accelerated tests to qualify a product to the customer's specifications
- Identification of reliability problems for corrective action
- Derating and redundancy decision making, based on tradeoffs between cost and risk
- Logistic planning, such as in maintainability decisions
- Measuring progress by monitoring reliability growth
- Warranty analysis

Failures of hardware are due to complex sets of interactions between (1) stresses and (2) material configuration of components, interconnections, and

TABLE 3.2 Material Failure Mechanisms

Overstress failure when any stress excursion exceeds strength	Wear-out failure when accumulated damage exceeds endurance
Performance failure not associated with material damage Mechanical Electrical Thermal Cosmetic Material failure mechanisms Fracture Buckling Yielding Interfacial fracture Electrical overstress Electrostatic discharge Dielectric breakdown Thermal breakdown	Material failure mechanisms • Fatigue • Creep • Metal migration • Corrosion • Wear • Aging Interdiffusion Depolymerizations Embrittlement

assemblies. A proper evaluation of reliability requires a systematic analysis of response of materials/configurations to stress. From the view of physics of failure, the characteristics of the failures of the product are defined by:

- failure mode,
- failure site, and
- failure mechanism.

3.6.1 Failure Mechanisms and Damage Models

Common failure mechanisms include those shown in Table 3.2.

3.6.2 Incorrect Mechanical Performance

Incorrect product response to mechanical overstress loads may compromise product performance, without necessarily causing irreversible material damage. Such a failure includes incorrect elastic deformation in response to mechanical static loads, incorrect transient response (such as natural frequency or damping) in response to dynamic loads, and incorrect time dependent (viscoelastic) response.

3.6.3 Incorrect Thermal Performance

Thermal performance failure can arise due to incorrect design of thermal paths in an assembly. This includes incorrect conductivity and surface emissivity of individual components, as well as incorrect convective paths for the heat-transfer path. Failure due to inadequate thermal design may be manifested as components running too hot or too cold, causing operational parameters to drift beyond specifications.

3.6.4 Incorrect Electrical Performance

Electrical performance failures can be caused by incorrect resistance, impedance, voltage, current, capacitance, or dielectric properties, or by inadequate shielding from electromagnetic interference, particle radiation, and electrostatic discharge. The failure mode can manifest as reversible drifts in electrical parameters and/or accompanying thermal malfunction. Here, only two major electrical design failures are discussed: failure caused by inadequate shielding from EMI and particle radiation.

3.6.5 Electromagnetic Interference

All electromagnetic waves consist of a magnetic (H) and an electrical (E) field. The relative magnitude of these fields depends on the nature of the emitter (source) and the proximity of the emitter to the shielding. The ratio of E to H is called the wave impedance.

When an electromagnetic wave encounters a discontinuity, such as metal shield, if the magnitude of the wave impedance differs greatly from the intrinsic impedance of the shield, most of the energy is reflected; very little is transmitted across the boundary and absorbed. Metals have an intrinsically low impedance because of their high conductivity. Therefore, for low impedance waves, less energy is reflected and more is absorbed because impedance of the metal shield more closely matches that of the wave.

3.6.6 Particle Radiation

The electrical failure modes caused by radiation are important to hardware design since they dictate, in part, the choice of packaging materials and allowable impurities in them. Radiation shielding may also be an important consideration in package design and configuration. Radiation effects on microelectronics may be a serious obstacle to further rapid increases in VLSI densities.

3.6.7 Yield

This is the first of the overstress material failures discussed in this chapter. Plastic deformations, caused by migration of microstructural defects (called dislocations) under mechanical loads in excess of yield strength (sometimes called flow stress) of the material, are irreversible. In other words, they manifest as a permanent deformation in material, even after the load is removed. Permanent deformations may be functionally inadmissible and be considered an overstress failure mechanism in some hardware. Common examples are overstress plastic strains in such precision structures as optic benches, metrological devices, and turbine blades.

3.6.8 Buckling

Buckling is an overstress failure mechanism caused by sudden catastrophic instability of a slender structure under applied compressive loads. Examples of buckling failures include lateral collapse of long slender columns under axial compression, bending-induced crippling of thin-walled structural beam sections, shear buckling of thin-walled tubular shafts under torsion, or wrinkling of thin plates and thin films under in-plane compressive and shear loads. Instability occurs when the compressive load reaches a critical threshold value, called the critical buckling stress. The critical buckling stress is a function of material properties (such as stiffness) as well as of structural geometry (such as slenderness ratio).

In mathematical terms, buckling is deformation along an unstable path orthogonal to the original deformation mode and can be solved by Eigen value or bifurcation theory. Postbuckling analysis utilizes large-deformation theory and can be accomplished through incremental nonlinear algorithms.

3.6.9 Fracture

Local microscale flaws, such as sharp microcracks, exist in most materials. Excessive stress concentrations at the tip of these sharp cracks can cause catastrophic propagation of the crack under overstress loads in brittle materials that exhibit little yielding and inelasticity before fracture. In ductile materials, a significant plastic zone may develop ahead of the crack tip due to localized yielding. The energy required to yield the material can increase the apparent resistance of a ductile material to fracture.

Designing for brittle fracture, a relatively new science, started during World War II because of persistent catastrophic fracture of the welded steel hulls of Allied Liberty ships, which became brittle in the cold Atlantic Ocean. Fracture is now recognized as a major cause of failure in engineering hardware, such as turbine blades, airframe parts, bridges, building frames, electronic dies, glass and ceramic components, and so on. Quasi-brittleness can lead to failure in hardened metal alloys and ceramics. Thermoset polymers can also undergo extensive microcracking and crazing due to brittle cracking. Brittle fracture can also occur due to the formation of brittle intermetallics in otherwise ductile materials, such as solder. A failure criterion based on stress is infeasible, because linear elastic analysis predicts infinite stresses at the tip of the flaw or crack, regardless of the magnitude of the far-field average or nominal stress. Hence, a new measure is required to quantify the severity of the stress field. This parameter, termed the *stress intensity factor*, indicates the intensity of the crack-tip stress field.

Hertzberg [28] postulated that catastrophic crack growth occurs when the energy required to create new free-crack surfaces in the fractured solid is less than the strain energy reduction in the solid due to changes in the crack length. The approach in fracture mechanics is to predict the level of far-field stress at which the crack will locally propagate.

The stress intensity factor, *K*, used to characterize the intensity of the crack-tip stress field, is defined in terms of the applied stress and the flaw size. For instance, in a plate of length *2h* and width *2b*, with a central crack of size *2a* such that a $<<$ *b* (indicating an infinite plate), *K* is:

$$K_{\mathrm{I}} = \sigma(\pi a)^{1/2} \tag{3.39}$$

where *a* is the applied far-field uniaxial stress. The critical or threshold value of the stress intensity factor, at which the crack will propagate, is a measure of the material's resistance to brittle fracture and is termed its *fracture toughness*. The fracture toughness depends on the orientation of the crack relative to the applied stress and is commonly characterized for three different fundamental fracture modes: crack-opening mode, shearing mode, and tearing mode. Fracture toughness values for common engineering materials are listed in ASM handbooks. The common design approach in fracture mechanics analysis is to compute the critical far-field load, based on the assumption that a characteristic flaw is located at the highest stressed region in the component. Details and illustrative examples may be found elsewhere.

Ductile fracture, like brittle fracture, is an overstress failure mechanism. It requires nonlinear modelling methods, because linear elastic theory of brittle fracture becomes inapplicable when there is large-scale plasticity at the crack tip. Ductile fracture can arise in many materials, such as aluminium, gold, copper, and solder, especially at high temperatures. Materials that behave in a brittle manner at relatively low temperatures and high strain rates can transition to ductile behavior at high temperatures and/or high deformation rates. The propagation of cracks in ductile materials requires higher energy, because the inelastic deformation at the crack-tip causes the material's apparent fracture toughness to increase.

3.6.10 Interfacial Deadhesion

Interfaces between dissimilar adhering materials can suffer adhesive failure under overstress loads. Examples include delaminations in composite materials and adhesion failures in bonded joints. Common examples in electronic packaging applications are failure at the interface of a die and the attach material, of a bond wire and the bond pad, and of the solder and the base material in a solder joint.

The interfacial strength depends on the chemical and mechanical properties of the interface. Interfacial adhesive failures can occur in diffusion-bonded, adhesively bonded, welded, soldered, and brazed joints between dissimilar adherents. One of the factors enhancing interfacial adhesive strength between two dissimilar materials is interdiffusion. However, dissimilar interdiffusion rates for the two adherent materials can degrade interfacial strength. Similarly, excessive intermetallic growth can cause a brittle interface of insufficient toughness.

3.6.11 Fatigue

Fatigue is the wear-out failure mechanisms. Cyclic mechanical deformations (or strains) and loads (or stresses) in a material can cause eventual failure, even though the peak strains may never exceed the ultimate ductility (strain at failure) of the material. Such failure is due to the accumulation of incremental damage with each load cycle, and is termed *fatigue*.

3.6.12 Creep

Some materials, such as thennoplastic polymers, solders, and many metals under mechanical stress at elevated temperature, can undergo a time-dependent deformation called creep. In reality, most deformations occur over a finite time period. For convenience in mechanics modelling, deformations that occur over very short time periods are treated as "instantaneous" and are termed *elastic* or *plastic*, depending on the reversibility of the deformation. Deformations requiring longer time periods are termed *creep*. Creep deformations are classified as viscoelastic (or anelastic) or viscoplastic, depending on whether the deformations are reversible.

Creep is a wear-out failure mechanism and can cause functional failure due to excessive deformation or act as a precursor to creep rupture. Creep occurs due to dislocation climb mechanisms, polymer chain reorientation, grain boundary sliding (superplasticity), intragranular void migration (self-diffusion), and/or intergranular or transgranular void migration (grain-boundary diffusion). Different creep mechanisms can dominate at different temperatures within the same material, and sometimes more than one creep mechanism can occur simultaneously. In many materials, there is a stage of decreasing creep rate (primary creep), followed by a stage of constant creep rate (secondary creep), and, finally, a stage of increasing creep rate (tertiary creep). The designer must ensure that, over the life of the package, creep strain is within design constraints.

3.6.13 Wear

Wear is a wear-out mechanism that is extremely important in all hardware that experiences impact from foreign particles or a sliding motion between surfaces in contact. For example, abrasive wear can occur due to continuous impingement of sand, water, or other foreign particles, causing gradual erosion; frictional wear can occur between gear teeth, sliding bearing surfaces, piston and cylinder assemblies, and so on, causing adhesive wear. In the case of liquid ducts, wear can occur due to liquid erosion of cooling ducts as a result of cavitation. Adhesive wear can lead to pitting and galling phenomena. Wear is not only a failure mechanism in itself, but also can leave hardware vulnerable to subsequent corrosion and overstress failure.

3.6.14 Aging Due to Interdiffusion

When two different materials are in intimate contact, molecules of one material can migrate into the other by diffusion or the ability of a material to migrate within a second material by atomic motion. From an atomic perspective, diffusion is the migration of atoms between lattice sites. The atoms must have sufficient energy to break bonds and reform them at another lattice site. The diffusion rate is a characteristic material property and can be measured in the laboratory.

Diffusion phenomena in themselves are not intrinsic failure mechanisms. For example, diffusion is a beneficial mechanism for forming diffusion bonds. However, diffusion can act as a failure agent when, for example, the diffusing medium is a harmful or corrosive chemical or when diffusion leads to microstructural aging, detrimental creep deformation, metal migration, and unbalanced interdiffusion. Interdiffusion is a time-dependent phenomenon and is therefore a wear-out failure mechanism.

3.6.15 Aging Due to Particle Radiation

Particle radiation is a common phenomenon in aerospace environments and in nuclear-power and particle-research establishments in terrestrial environments. Radiation damage includes both mechanical and electrical failures. The mechanical failure mechanism is typically an embrittlement aging phenomenon of the wear-out type. Common examples are exposed hardware in space satellites, reactor vessels, and such. The electrical phenomenon is an overstress phenomenon that causes "soft errors" due to the passage of single radiation particles through LSI/VLSI electronic hardware.

Radiation damage causes different aging in types of different materials. Radiation damage is a time-dependent wear-out phenomenon, and is of concern in metallic, ceramic, and polymeric materials. In metals and ceramics, radiation causes point defects, such as pairwise combinations of vacancies and interstitial atoms (Schottky defects), by knocking atoms out of molecular lattice structures and lodging them in interstitial sites. These point defects cause embrittlement aging, which can be countered by annealing operations. More importantly in electronic packaging applications, these defects can also alter thermal, optical, and electrical properties, impairing the operation of active devices. In polymeric materials, radiation aging is caused by breaks in polymer chains or changes in the degree of polymerization due to chain branching. Either of these can reduce the strength of the polymer. In its most common form, this can lead to photo degradation of polymers under prolonged exposure to UV radiation in strong sunlight. Stabilizers are sometimes needed to combat such wear-out failures.

3.6.16 Other Forms of Aging

There are a variety of other forms of aging that can alter a material's performance overtime. Examples include hydrogen embrittlement, thermally induced depolymerization, increased cross-linking leading to embrittlement in thermosetting polymers, and grain growth in crystalline materials. Detailed discussions of all these mechanisms are, however, beyond the scope of the present discussion.

3.6.17 Corrosion

Corrosion results from the chemical or electrochemical degradation of metals. Corrosion is a time-dependent wear-out failure mechanism and can act as a precursor either to subsequent overstress failure by brittle fracture or to subsequent wear-out failure by fatigue-crack propagation. Corrosion can also alter the electrical and thermal behavior of materials in the microscale. The three most common forms of corrosion are uniform, galvanic, and pitting corrosion. The corrosion reaction rate depends on the material, the presence of an electrolyte, the presence of ionic contaminants, geometric factors, and local electrical bias.

Uniform corrosion is a chemical reaction occurring at the metal–electrolyte interface uniformly throughout the surface. The continuation of the corrosion process and its rate depend on the nature of the corrosion product. If the corrosion product is soluble in the electrolyte (say, water), it can be dissolved away, exposing fresh metal for further corrosion. On the other hand, if the corrosion product forms an insoluble, nonporous, adherent layer, it limits the rate of reaction and finally stalls the corrosion process.

Galvanic corrosion occurs when two or more different metals are in contact. Each metal is associated with a unique electrochemical potential. When two metals are in contact, the metal with the higher electrochemical potential becomes the cathode, and the other metal, the anode. The electrical contact between dissimilar metals leads to the formation of a galvanic cell. The rate of galvanic corrosion is governed by the rate of ionization at the anode (i.e., the rate at which anode material passes into solution), and this, in turn, depends on the difference in electrochemical potential between the contacting two metals. The conductivity of the corrosion medium affects both the rate and distribution of galvanic attack. In solutions of high conductivity, the corrosion of the more active alloy is dispersed over a relatively large area. In solutions with low conductivity, most of the galvanic attack occurs near the point of electrical contact between the dissimilar metals.

Pitting corrosion occurs at localized areas, causing the formation of pits. The corrosion conditions produced inside the pit accelerate the corrosion process. As the positive ions at the anode go into solution, they become hydrolyzed, producing hydrogen ions in the process. This increase in acidity in the pit destroys the adhering corrosion products, exposing more fresh metal to attack. Since the oxygen availability in the pit is low, the cathodic reduction reaction can occur only at the mouth of the pit, thus limiting lateral growth of the pit.

Surface oxidation, another common type of corrosion in metallic materials, is governed by the free energy of formation of the oxide. For example, there is a large driving force for the oxidation of aluminium and magnesium but much less of a force for copper, chromium, and nickel. Depending on the stoichiometry of the corrosion reaction, the type of the oxide formed can be either porous or dense-packed. The oxide type frequently governs the subsequent rate of corrosion. For instance, a thick, nonporous oxide layer may act as a protective barrier and inhibit further corrosion by cutting off the oxygen supply to the surface, as with aluminium and stainless steel. Sometimes, the volume of the corrosion product (oxide) may be so much higher than the base material that the oxide layer peels off. Such scaling failure exposes the underlying metal to fresh attack.

Corrosion is a leading cause of damage and failure in engineering hardware, and prevention, cure, and replacement costs are significant.

3.6.18 Metal Migration

This wear-out mechanism, important in electronic hardware, is driven by diffusion phenomena. There are many types of metal migration, including electromigration, cathodic dendritic growth, and conductive anodic filament (CAF) growth. Dendritic growth is essentially an electrolytic process in which the metal from the anode region migrates to cathodic areas. Metal migration leads to an increase in leakage current between the bridged regions or causes a short if complete bridging occurs (migrative resistance shorts). Although Ag migration has been most widely reported, depending on environmental conditions, many other electronic metals, like Pb, Sn, Ni, Au, and Cu, can also migrate. Being time-dependent, this is a wear-out mechanism.

Metal migration is governed by the availability of metal, the presence of electrolytes, such as condensed water and ionic species, and the existence of a voltage differential. Metals known to be susceptible to metal migration should be protected from water vapor and ionic contamination. As the migration phenomenon is an electrolytic process, it is essential to have a conducting medium. Ionic species include impurities, such as chlorides or products generated during corrosion. The driving force necessary to cause metal migration is the potential difference existing when the electronic hardware is in a biased condition. While the primary stress for this failure mechanism is an electrical potential gradient, it can be accelerated by secondary stresses, such as moisture, ionic contaminants, and temperature.

3.7 MAINTAINABILITY ANALYSIS AND DESIGN

Attention was given to maintainability features in the equipment design stage in the late 1950s. By 1959, the USAF had published MIL-STD-26512, a document focusing on maintainability. Since then several standards and military documents on maintainability have been developed.

3.7.1 Definitions Used in Maintainability Engineering

Maintainability: The probability that an inoperable item (i.e., failed equipment) is restored to a working state within a defined downtime period.

Availability: The probability that at time *t* the system is functioning normally when used under specified conditions, where administrative time, logistic time, and operating time are the components of total time.

Down time: The total time during which the equipment is in an unsatisfactory operable state. MDT = active maintenance time + logistic delay + administration delay.

Operating time: The time during which the equipment is operating satisfactorily, as expected by the operator, although the equipment's unacceptable operation is sometimes due to the maintenance person's assessment.

Active repair time: The equipment's downtime in which one or more maintenance people are performing their duty to effect an equipment repair (Fig. 3.10).

Preventive maintenance: Maintenance that is carried out to retain equipment in an acceptable operating state by providing orderly detection and inspection in addition to prevention of incipient failures.

Corrective maintenance: Maintenance that is carried out to restore failed equipment to acceptable operable conditions.

3.7.2 Measurements

MTBM: mean time between maintenance
MTTR: mean time to repair
MAMT: mean active maintenance time
MPMT: mean preventive maintenance time
MDT: maintenance down time

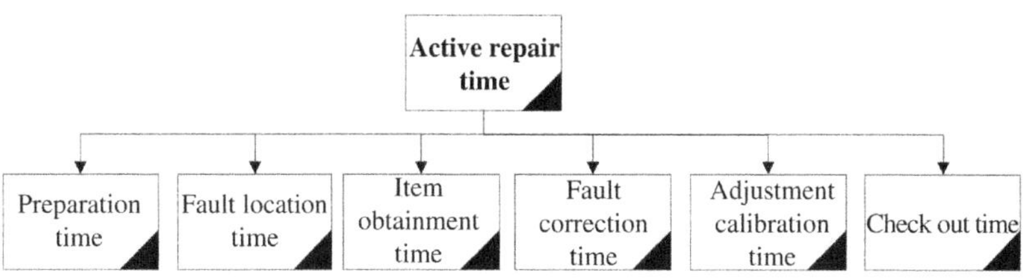

FIGURE 3.10 Components of active repair time.

LDT: logistic delay time
ADT: administration delay time
Cost/OH: maintenance cost per system operating hour
MMH/OH: mean man-hours per operational hours

3.7.3 Maintainability Function

Where t is the variable repair time, $f(t)$ density function:

$$M(t) = \int_0^t f(t)\,dt \tag{3.40}$$

3.7.4 Often Used Maintainability Distributions

Exponential:

$$M(t) = \int_0^t ue^{-ut}\,dt = 1 - e^{-ut} \tag{3.41}$$

Where u is repair rate, $MTTR = 1/u$. The time to repair electronic equipment follows this distribution.

Log normal distribution:

$$\begin{aligned} M(t) &= \frac{1}{\sigma_1\sqrt{2\pi}}\int_0^t \exp\left[-\frac{(\ln t - \mu_1)^2}{2\sigma_1^2}\right]dt \\ &= \Phi\left(\frac{\ln t - \mu_1}{\sigma_1}\right) \end{aligned} \tag{3.42}$$

Aircraft: time to repair follows log-normal distribution.

Repair rate function:

$$\mu(t) = \frac{f(t)}{1 - M(t)} \tag{3.43}$$

3.7.5 Availability Models

Inherent availability model:

$$A_i = \frac{\text{MTBF}}{\text{MTBF} + \text{MCT}} \tag{3.44}$$

Excludes preventive, scheduled maintenance actions, logistic delay time, and administrative delay time.

Achieved availability:

$$A_a = \frac{\text{MTBM}}{\text{MTBM} + \text{MAMT}} \tag{3.45}$$

MAMT: Mean active maintenance time, excludes logistic and administrative delay time.

Operational availability:

$$A_o = \frac{\text{MTBM}}{\text{MTBM} + \text{MDT}} \tag{3.46}$$

MDT: Mean maintenance down time; MTBM: Mean time between maintenance.

System instantaneous availability:

$$Av(t) = \frac{k}{D} + \frac{\lambda}{D} e^{-Dt} \tag{3.47}$$

λ, k are system constant failure rate and repair rate, respectively, $D = k + \lambda$.

System mission availability:

$$Av_{\text{m}}(t) = \left(t_{\text{b}} - t_{\text{a}}\right)^{-1} \int_{t_{\text{a}}}^{t_{\text{b}}} Av(t)\,dt \tag{3.48}$$

If $t_{\text{a}} = 0$ and $t_{\text{b}} = t$, then

$$Av_{\text{m}}(t) = \frac{k}{D} + \frac{\lambda}{D^2 t}\left(1 - e^{-Dt}\right) \tag{3.49}$$

3.7.6 Effectiveness Models

Definition: probability that an item is able to meet an operational requirement within a specified period when used under a stated condition.

$$P = P_{\text{r}} \cdot R_{\text{m}} \cdot P_{\text{d}} \tag{3.50}$$

where P, system's effective probability; P_{r}, probability of operational readiness; R_{m}, mission's reliability; P_{d}, probability of design adequacy.

3.8 SPECIFICATION OF MAINTAINABILITY [29]

A maintainability specification's form can vary considerably. It may be a statement defining the required test program to meet purchaser's requirements, or a formal specification defining complete management structure and setting standards for all aspects from operating environment demands for life testing, to detailed definitions of environmental or functional testing.

Maintainability clauses normally contain three elements as follows:

- The objective or required value of the relevant maintainability characteristics, expressed in performance terms.
- The conditions of use, storage and maintenance, and life of item, during which this maintainability is required.
- The means by which the required maintainability is to be, or has been assured.

3.8.1 Quantitative Maintainability Clauses

Complete statement of maintainability requirements will cover four broad areas, as follows:

1. Maintainability characteristics to be achieved by the item design.
2. Constraints to be placed on item deployment that will affect its maintenance.
3. Maintainability program requirements to be accomplished by the supplier to assure the delivered item have the required maintainability characteristics.
4. Provision of maintenance supports.

3.8.2 Qualitative Maintainability Requirements

Qualitative approach considers specification of design disciplines and the degree to which the item concurs with a specific maintenance and support policy, where the quantitative requirement contains numerical values. Such policy could include statements, such as the following:

- Repair shall be performed by personnel of stated skill level
- Repair should be performed by replacement of recoverable units
- Replaceable parts shall be plug-in units
- Maintenance shall be performed according to defined and established procedures
- Failed parts isolation shall be performed by built-in test equipment for 95% of all cases

Examples of qualitative aspect for which requirements may be specified are:

- Maintenance skill level requirements
- Need for special tools or test equipment
- Need for adjustments
- Parts standardization
- Clear subsystem function identification
- Visual inspection access
- Built-in test facilities
- Properly marked test points
- Color coding and labels as appropriate
- Use of plug-in units

- Use of captive fasteners
- Use of handles on replaceable units
- Scope and range of technical manuals; and
- Human factor limitations in design of the item

3.8.3 Choice of a Maintainability Characteristic

Quantitative maintainability characteristics are used to express maintainability numerically. Maintainability is generally approached from the standpoint of returning equipment to an operating condition following failure (corrective maintenance) or keeping the system from failing (preventive maintenance). The most common objective is related to the time an item is in a nonoperative status due to maintenance.

Active repair time is often used to specify maintainability and include the following subelements:

- Diagnosis (failure detection, localization of cause, etc.)
- Technical delays (typical technical delays include setting time, cooling, interpretation and application of information, interpretation of display, readout, etc.)
- Restoration (disassembly, interchange, reassembly, alignment, etc.)
- Final check (testing procedure as necessary)

Characteristics: A variety of other maintainability characteristics may be specified for the item. Their characteristics are:

- Active maintenance time (mean, median, maximum)
- Active corrective time (mean, median, maximum)
- Routing inspection interval
- Maintenance cost per operating hour (mean)
- Numbers of hours labor per operating hour (mean)
- Number of personnel per maintenance action (mean); and
- Maintenance support cost for the life cycle (cost)

Example:

1. The mean time to repair (MTTR) at intermediate level shall be *X* minutes. *Y*% of all maintenance tasks shall be completed in less than *Z* minutes.
2. Preventive maintenance time shall not be required.
3. Maintenance reliability, the probability an equipment is capable of performing its functions following a satisfactory maintenance checkout, shall be greater than *X*%.
4. All operator level maintenance tasks shall be completed in less than *Y* minutes without use of special tools.

Required value of the maintainability characteristics: In some cases, two maintainability characteristics' values may be specified, which will better determine the distribution. For example, as well as specifying an MTTR for equipment, the maximum time to repair (i.e., the longest repair time) may be specified.

3.9 ASSESSMENT AND PREDICTION OF MAINTAINABILITY

Maintainability assessment is the process by which quantitative values are assigned to reliability and maintainability.

Maintainability assessment is required when:

- Establishing the maintainability required of a product;
- Predicting the maintainability of a product still in design, development, or premanufacturing stage; and
- Establishing whether a product in service has performed, or is performing, in such a way as to satisfy the specified value of the reliability characteristic, and whether it is likely to continue to do so for the rest of its design life.

3.9.1 Maintainability Prediction [30]

Maintainability prediction is the estimation of maintenance workload (preventive and corrective) associated with the proposed design. Maintainability prediction should be accomplished immediately following definition of the basic system. This is the earliest time when sufficient data is available to perform a meaningful quantitative evaluation of design characteristics in terms of performance and maintenance. At this early stage of system design process, maintainability predictions can still influence the design approach. As the system design progresses to the detailed level, more complete design information becomes available and consequently estimation of system maintainability characteristics becomes more accurate. The estimate should be updated continuously as the design progresses to provide visibility needed to ensure the specified requirements have a high probability achievement. Predictions are applicable to all programs and all system and equipment types. However, they are particularly pertinent in programs where risks are high or unknown, and failing to achieve the maintainability requirements, highly undesirable.

3.9.2 Prediction Advantages

A significant advantage of using maintainability prediction is it highlights for the designer, areas of poor maintainability that justify product improvement, modification, or a change of design. Another useful maintainability prediction feature is that it allows the user to make an early assessment of whether predicted downtime, personnel quality and quantity, tools and test equipment are adequate and consistent with the needs of the system's operational requirements.

3.9.3 Techniques

The effectiveness of maintainability prediction as an evaluation tool depends on the technique and accuracy of input data. This in turn is based on the applied knowledge and insight of the analyst. Presently several maintainability prediction techniques are used. Procedures vary depending on the specific need for measurement, differences in imposed requirements, peculiarities of equipment being measured, and individual or company preferences.

3.9.4 Basic Assumptions and Interpretations

Every maintainability prediction procedure depends upon use of recorded reliability and maintainability data and experience obtained from comparable systems and components under similar use and operation conditions. It is also customary to assume the principle of transferability. This assumes data accumulating from one system can be used to predict maintainability of a comparable system undergoing design, development, or study. This procedure is justifiable when the required degree of commonality between systems can be established. Usually during the early design phase, commonality can be only broadly inferred. As the design is refined, commonality is extendable if a high positive correlation in equipment functions, maintenance task times, and levels of maintenance is established. History has shown the advantages greatly outweigh the burden of making a maintainability prediction.

3.9.5 Elements of Maintainability Prediction Techniques

Each maintainability prediction technique utilizes procedures specifically designed to satisfy its application method. All maintainability prediction methods are dependent on at least two basic parameters:

1. Failure rates of components at the specific assembly level
2. Repair time required at the maintenance level involved
   a. *Failure rates*: Many sources record failure rates of parts as a function of use and environment. Failure rates are used in maintainability prediction to provide an estimate of the relative frequency of failure of components in the design. Similarly, relative frequency of component failure at other maintainable levels can be determined by employing standard reliability prediction techniques using parts failure rates. Another use of failure rates is to weight repair times for various categories of repair activity, to provide an estimate of its contribution to total maintenance time.
   b. *Repair times*: Repair times are determined from prior experience, simulation of repair tasks, or data secured from similar applications. Most procedures break maintenance action into a number of basic tasks whose time of performance is summed to obtain total time for the maintenance action.

3.10 MAINTAINABILITY DESIGN: THE AFFECTED FACTORS

- Fault diagnosis: BIT technique
- Fault isolation
- System complexity
- Accessibility
- Maintenance environment
- Packaging
- Handling of tools and other materials
- Limiting clearance
- Standardization and interchange ability
- General criteria

3.11 MAINTAINABILITY DESIGN: CRITERIA

- Providing adequate accessibility, workplace, and work clearance
- Reducing the need for, and frequency of, maintenance activities
- Reducing maintenance downtime
- Reducing maintenance support costs
- Reducing maintenance personal requirements
- Reducing potential for maintenance error
- Providing a built-in test capability

3.12 MAINTAINABILITY DESIGN: ALLOCATION

Purpose: Allocate the system's maintainability requirement to subsystem
Methods: Weighting factors
Assumption: Failure rate follows exponential distribution. Repair rate follows log-normal distribution. λ_s, system failure rate; λ_I, subsystem failure rate; *n*, the number of affected factors; *m*, the number of subsystems; W_i, the average of weighting factors of each subsystem.

$$\lambda_s = \sum_{i=1}^{m} \lambda_i \tag{3.51}$$

$$t_i = \frac{w_i}{w} MTTR \tag{3.52}$$

$$w = \frac{\sum_{i=1}^{m} \lambda_i w_i}{\lambda_s} \tag{3.53}$$

In a series system with four subsystems, with failure rates of each subsystem $\lambda_1 = 0.001$, $\lambda_2 = 0.002$, $\lambda_3 = 0.003$, and $\lambda_4 = 0.004$, respectively, and with the

TABLE 3.3 Weighting Factors of Subsystems

Subsystem, $m = 4; i = 1, 2 \dots m$	Weighting factors (total 10 for each factor), $i = 1, 2 \dots 6$						Wi $n = 6; i = 1, 2 \dots n$
	1 Fault isolation	2 Complexity of the system	3 Environmental	4 Packaging	5 Handling	6 Accessibility	
1	7	3	2	1	3	2	3.16
2	2	3	1	2	1	4	2.167
3	1	4	3	3	1	1	2.167
4	0	0	4	4	5	3	2.66
$W = 2.538$							

required system MTTR of 2 h, allocate the requirement to subsystems as shown in Table 3.3.

The allocated MTTR of subsystems: $T_1 = 2.49$, $T_2 = 1.70$, $T_3 = 1.70$, $T_4 = 2.096$.

3.13 MAINTAINABILITY DESIGN—LIMITING CLEARANCE [31]

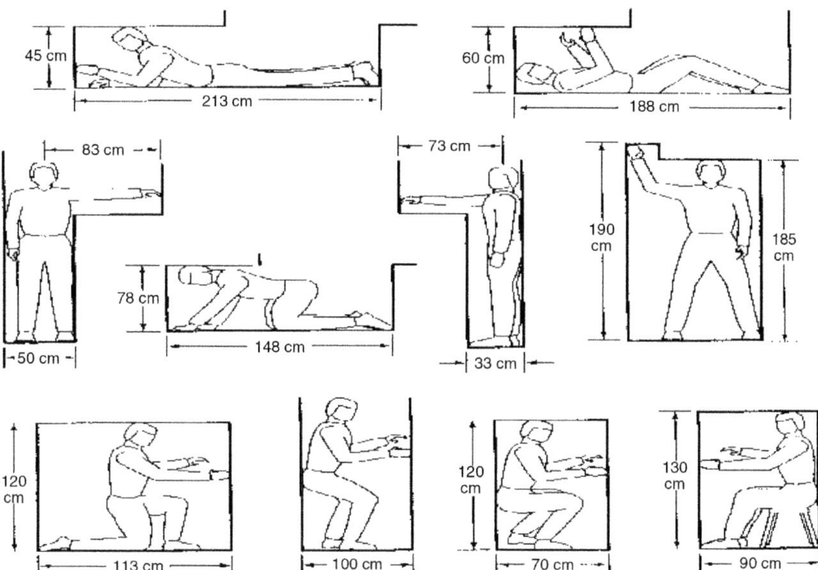

3.14 MAINTAINABILITY DESIGN—ACCESSIBILITY

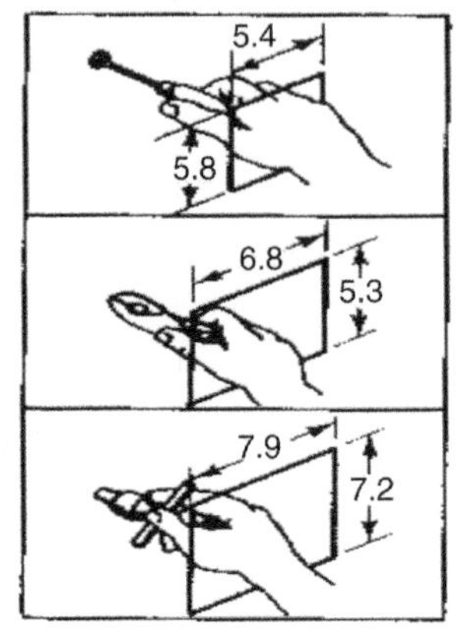

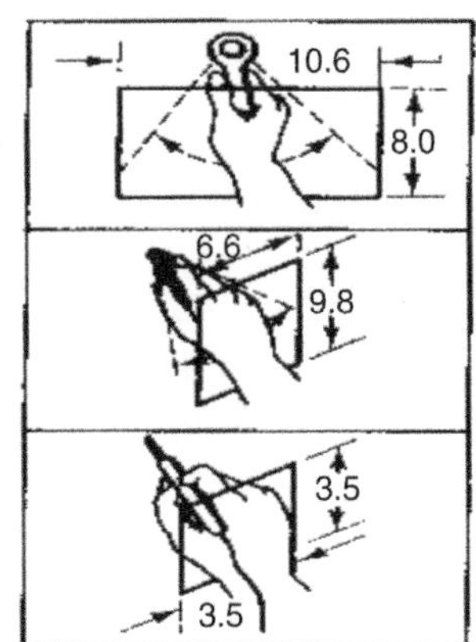

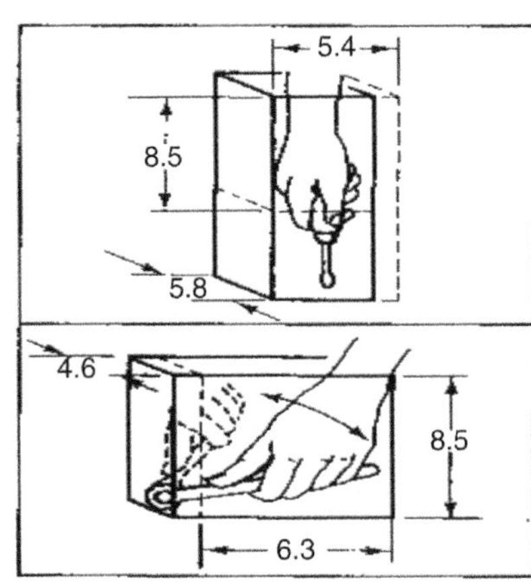

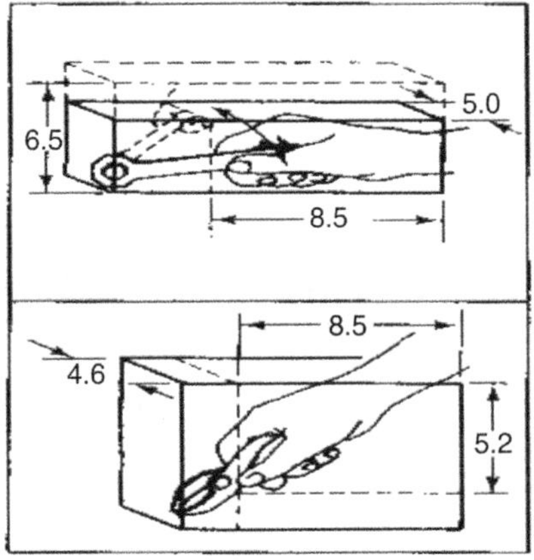

3.15 MAINTAINABILITY DESIGN—PACKAGING

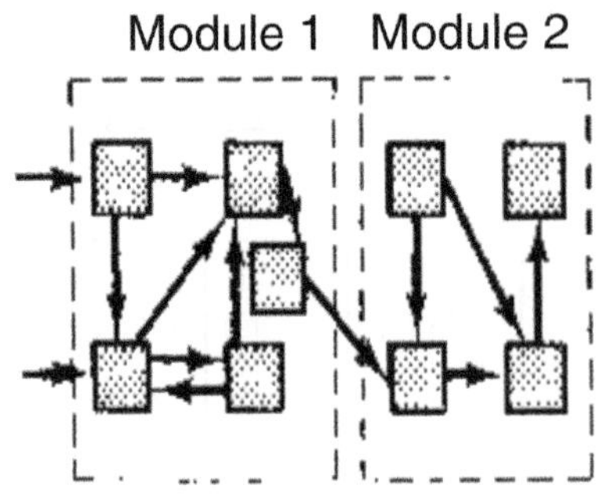

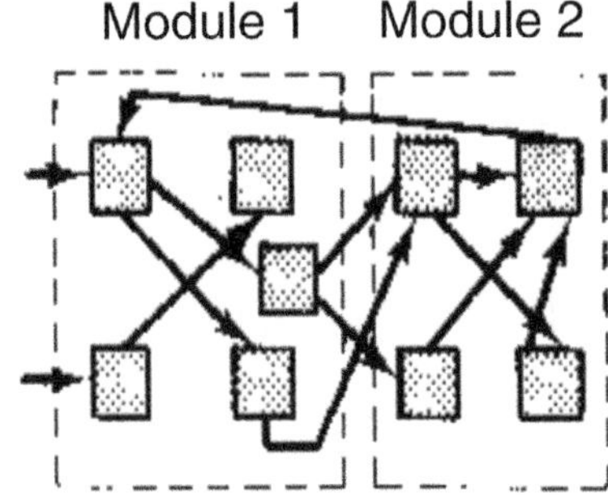

3.16 MAINTAINABILITY DESIGN—STANDARDIZATION AND INTERCHANGE ABILITY

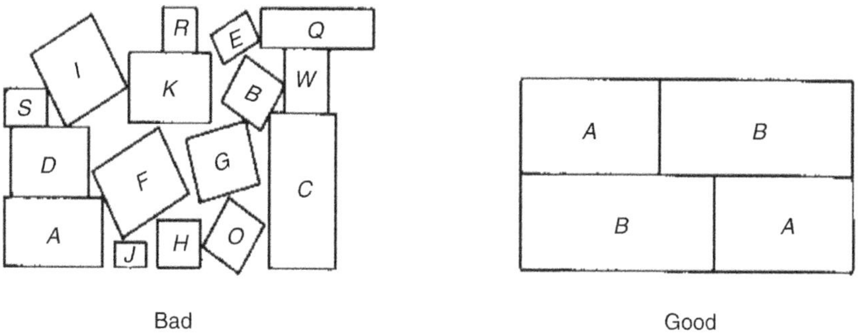

3.17 MAINTAINABILITY DESIGN—INSTALLATION-COMPONENTS ARRANGEMENT

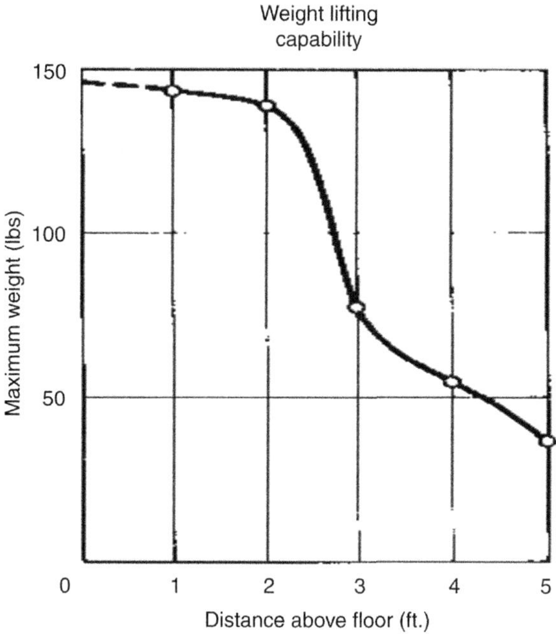

3.18 MAINTAINABILITY DESIGN—GENERAL CRITERIA

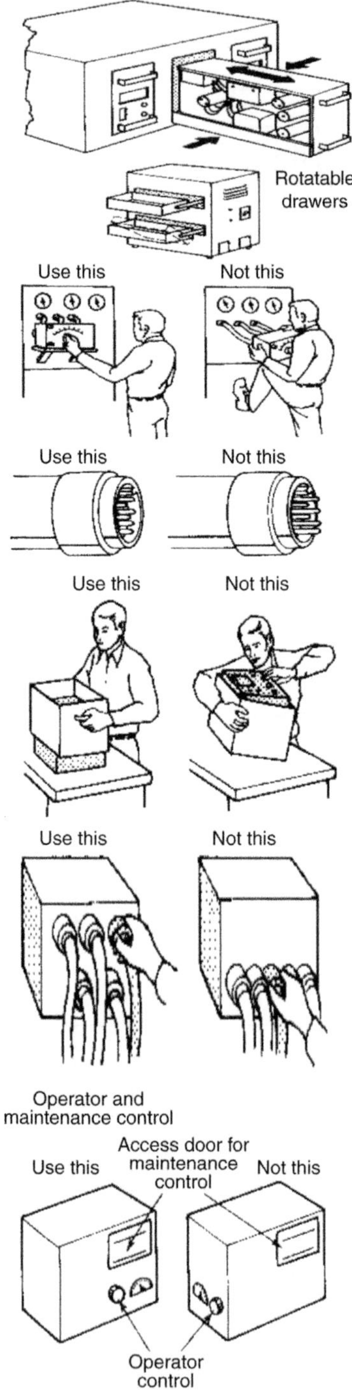

3.19 MAINTAINABILITY DEMONSTRATION AND TESTING [32]

Maintainability specifications written into a contract are in effect only targets or goals, unless there is an actual assessment of the maintainability parameters of the developed system/equipment.

The primary function of maintainability test and demonstration is to "verify maintainability" that has been "designed-in" and "built-in" to the system/equipment. Up to this point in development, the tasks of the maintainability program have been analytical in nature, providing a confidence that both quantitative and qualitative maintainability requirements would be met.

3.19.1 Maintainability Testing Program

There can be three phases to a maintainability testing program.

Maintainability verification is conducted incrementally during development on mock-up models and early hardware designs, with the intention of providing progressive assurance that maintainability requirements can be achieved and earlier modelling and allocation were accurate.

Maintainability demonstration occurs at the end of development, to determine whether contractual requirements have been achieved.

The demonstration is performed on as close-to-production hardware as possible (i.e., final prototype or preproduction item), conducted in an environment that simulates, as closely as possible, the operational and maintenance environment specified for the item. The environment should be representative of the working conditions, tools, support equipment, repair parts, facilities, and technical publications required during operational service.

Maintainability evaluation occurs in the field environment. Its objectives are to evaluate the impact of the actual operation, maintenance and support environment on the maintainability, to evaluate correction of deficiencies detected during the maintainability demonstration, and to demonstrate depot level maintenance tasks when applicable. All evaluation items should be production or production equivalent items.

3.19.2 Maintainability Demonstration

To fulfill maintainability demonstration requirements for a typical program, a contractor is obligated to demonstrate that equipment meets specified maintainability requirements. The accomplishment of such a demonstration in a realistic operational environment is often impractical. In certain instances demonstrations can be accomplished in an environment closely approximating a true operational situation. In other words, a contractor demonstration may be conducted at the customer's facility, employing customer personnel, on equipment installed and ready for operational use. However, such demonstrations are generally scheduled at specific times and faults are simulated in the equipment to simulate maintenance requirements. Although this type of situation does not completely reflect normal user operations (since failures are induced and subsequent demonstrations planned, eliminating some of the randomness normally involved), it can provide a close simulation.

3.19.3 Test Conditions

Test conditions for formal maintainability demonstrations include:

- Maintainability requirements;
- Maintenance policy;
- Demonstration model configuration;
- Test environment;
- Test personnel;
- Technical data;
- Support equipment; and
- Spare parts.

3.19.4 Maintenance Task Selection

The assurance that the proposed demonstration reflects total system maintainability depends on the maintenance task selection process. This process involves identification of a representative sample (based on the expected percentage contribution to total maintenance requirements) of maintenance tasks to be demonstrated. The process does not include actual random (unplanned) failures occurring during the test, but it does include a variety of induced failures to ensure adequate coverage.

3.20 MAINTAINABILITY AND RELIABILITY PROGRAM ACTIVITIES DURING THE PHASES OF A PROJECT [33]

3.20.1 Definition Phase

- Feasibility study
- Statement of reliability and maintainability
- Objectives and requirements
- Reliability and maintainability specification and contract formulation

3.20.2 Design and Development (Including Initial Manufacture)

- Analysis of parts, materials, and processes
- Analysis of established and novel features
- Failure mode, effect, and criticality analysis
- Incident sequence analysis (fault tree analysis)

- Stress and worst case analysis
- Reliability prediction
- Redundancy analysis
- Human factors
- Design review
- Design audit
- Design change control
- Maintainability analysis
- Maintainability prediction
- Maintainability design criteria
- Safety program
- Test plans
- Parts and subassembly testing
- Performance and environmental testing
- Accelerated testing
- Endurance testing
- Reliability growth testing
- Development reliability demonstration testing
- Maintainability test and demonstration
- Data collection, analysis, and feedback

3.20.3 Production

- Preservation of reliability achievement
- Quality conformance verification
- Screening (run-in, bed-in, or burn-in) of components and assemblies
- Production reliability demonstration testing
- Maintainability in production
- Additional software check

3.20.4 Installation and Commissioning

- System acceptance
- Commissioning tests
- Reliability growth
- Reliability and maintainability demonstration
- Data collection
- Reliability and maintainability assessment

3.20.5 Operation-Usage and Maintenance

- Data collection, analysis, feedback, and redesign
- Modification
- Maintenance

TABLE 3.4 Responsibility Interface of Maintainability and Maintenance

Maintainability	Maintenance
• Requirements • Design criteria • Repair policy • Test philosophy • Cost analysis	• Concepts • Procedures • Personnel skills • Training • Support equipment • Provision

3.21 MAINTAINABILITY MANAGEMENT

3.21.1 Responsibilities Interface of Maintainability and Maintenance [34]

The responsibilities interface of maintainability and maintenance is given in Table 3.4.

3.21.2 Maintainability Analysis

- Preparation of maintainability demonstration documents
- Trade-off analysis of maintainability
- Allocation of the maintainability
- Reviewing system requirement
- Prediction of system maintainability
- Participating in the engineering maintenance analysis

3.21.3 Maintainability Design

- Preparing design report
- Taking part in item design review
- Monitor item design
- Approving product design drawing
- Consulting service

3.21.4 Maintainability Administration

- Organizing
- Staffing
- Preparing maintainability program plan
- Budgeting and scheduling
- Developing procedure and policies
- Design reviewing
- Coordination
- Documentation

Chapter 4

RCM and Integrated Logistic Support

Chapter Outline

4.1 INTRODUCTION

This chapter aims to help the reader develop an understanding of the methodology behind reliability-centered maintenance (RCM) analysis, including broad logistic support analysis, processes, and how the outcome contributes to effective maintenance management. In developing the maintenance plan, techniques of logistic support analysis (LSA) are used to determine the requirements to support a specified operational schedule or readiness profile. These requirements were established on the FMECA data derived from failure patterns and the criticality of effects. In following up this analysis

Reliability Based Aircraft Maintenance Optimization and Applications
http://dx.doi.org/10.1016/B978-0-12-812668-4.00004-6

maintenance intervals, task types, and the performance levels are determined. This is followed by decisions on associated procurement tasks, stocking levels for spare parts, and other items needed to support these plans. Overall the process is known as integrated logistic support (ILS) management. Optimization of maintenance efforts in support of the operations pattern will be discussed as well.

4.2 MAINTENANCE ANALYSIS PROCEDURES

The philosophy for determining maintenance requirements for an aircraft type has developed progressively over recent decades. Until the early 1970s requirements for maintenance tasks for both civil and military aircraft were generally aligned to a safe-life concept of airworthiness. Any failure was unacceptable. Fatigue-based cracking was avoided by factoring the life back from the hours at which a failure was expected to occur. Lack of structural redundancy and the lack of "fail–safe" concepts in design often required rather arbitrary and relatively short fatigue lives for some wing and fuselage structures.

Aircraft were inspected at regular "hard time" intervals based on past practice for similar generic types. Major inspections of aircraft required a high degree of disassembly and structural inspections were expected to look at virtually the whole aircraft.

Most components, particularly vital ones, such as engines, would have an overhaul life set during the prototype development testing. Most maintenance was determined by a wear-out assessment, items being inspected, removed for overhaul, or bay-serviced according to their flying hour history at set periods or intervals.

These "hard-time" intervals might be extended or perhaps reduced on some basis of experience as approved by the regulatory authority. This process was reasonably thorough and secure as Friend [27] points out: "the procedure was conservative; most components started with low approved lives and their reliability had to be established by trial in service." But with the benefit of hindsight it was grossly inefficient, likely to induce faults in the overservicing work done and causing significant economic inroads into availability.

4.2.1 The MSG Series Procedures

In civil aviation the concepts of on-condition removal were introduced with a FAA Industry Reliability Program that led to the adoption by its Maintenance Steering Group of a maintenance analysis process called MSG-1, first applied to the Boeing 747 in 1968. MSG-2 followed in 1970 and included condition monitoring. It was used with the DC-10 and Lockheed L1011. The Association of European Airlines developed a similar process called EMSG in 1972 used on Concorde and the Airbus A300. Ultimately the MSG-3 approach was developed and published in 1993 by the United States and Europeans jointly and applied to Boeing 757 and 767 aircraft. In its updated form this remains the current standard approach for airline aircraft [27].

As part of the civil certification process, a representative group of people, known as the Maintenance Review Board (MRB) from the manufacturer, regulator, and operational airlines meet in working groups to define basic maintenance requirements for the type. This is then documented and issued with the regulator's approval. Each airline may develop its own maintenance schedules to meet these basic requirements but varied to suit particular local airline requirements and these schedules in turn are approved by the regulatory authority and are legally enforceable.

4.2.2 Reliability-Centered Maintenance

Reliability-centered maintenance (RCM) for civil aviation was developed by United Airlines in the United States in the mid-1980s. It corrected a number of inadequacies in the logic and scope of MSG-2 including aspects of fatigue design rules, electronic control and display systems, and the impact of prices during the oil crisis on airline economics. The logic of MSG-2 was to build upward from the component to the systems level, while MSG-3 works from the top–down, from the impact of any unreliability to maintenance tasks required to preserve this reliability.

The RCM concept also had a major influence on military aircraft maintenance planning detailed further later. Royal Australian Air Force (RAAF) doctrine requires the acceptance of the following:

- Design of equipment establishes the consequences of failure.
- Redundancy can overcome problems of safety consequences.
- Scheduled maintenance can prevent or reduce frequency of functional failures.
- Scheduled maintenance can achieve the inherent reliability of an item but cannot improve it.
- Reliability problems cannot be solved by scheduled maintenances.
- On-condition inspection is the most effective tool of preventative maintenance by making it possible to prevent functional failures.
- Scheduled maintenance programs must adapt to changing needs by responding to service experience.
- Product improvement is a normal part of the development cycle for all new equipment.

4.2.3 MSG-3 Logic

The logic process is divided into two independent logic paths.

Structures based on guidelines of FAR 25,571 for damage tolerant design of structurally significant items (SSIs). The equivalent US military requirements are set out in Mil-STD-1530 A (1995).

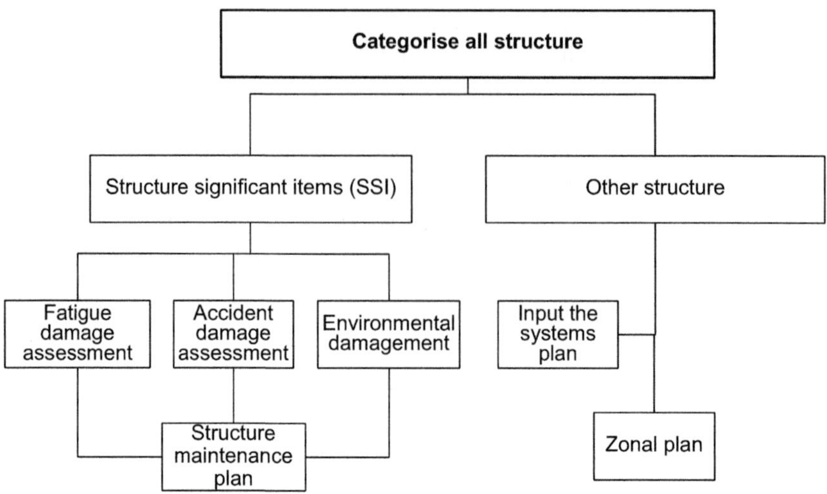

FIGURE 4.1 **Simplified MSG-3 structural logic diagram.**

Systems and power plants that consider maintenance significant items (MSIs) are sometimes (e.g., RAAF) called functionally significant items (FSIs), defined as those of which a failure:

1. Could affect safety in the air or on the ground, and/or
2. Is undetectable during operations and/or
3. Could have significant operational economic impact, and/or
4. Could have significant nonoperational economic impact

4.2.4 Structures

Items are evaluated for their significance, susceptibility to fatigue, environmental or accidental damage, and those identified will be assigned preventative maintenance tasks. Remaining items are monitored by "zonal checks", that is, a general visual survey of a defined area or volume. See Fig. 4.1 for an illustration of the simplified logic diagram.

The process requires a worksheet recording all descriptive information, input data, and decisions made on tasks and intervals and so forth.

An SSI is defined as a detail, element, or assembly that is judged significant due to the reduction in aircraft residual strength or loss of structural function that results from its failure. Other structures, not judged to be SSIs, are defined within zonal boundaries.

Once the analysis process has defined and described the item, determined its significance, and collated the manufacturer's design, material, and manufacturing data, the logic steps aforementioned are followed. SSI's are classified as either:

Damage tolerant: That is, "it can sustain damage and the remaining structure can withstand reasonable loads without structural failure or excessive structural deformation until the damage is detected."
Safe life: That is, "Structure that is not practical to design or qualify as damage tolerant. Its reliability is protected by discard limits that remove items from service before failures are expected."

Rating values are associated with various effects on a SSI to rank their importance and contribute to the judgement on inspection methods and intervals.

4.2.5 Fatigue Damage

In relation to fatigue damage, the following rating issues are considered:

- Threshold for detectable size fatigue damage
- Detection standards for applicable types of inspections
- Crack growth assessment for repeat inspection interval
- Residual strength damage size
- Fleet size/usage assessment
- Target values for scheduled maintenance check intervals.

4.2.6 Environmental Deterioration

Environmental deterioration may or may not be time dependent and is influenced by the effectiveness and durability of surface protection schemes. The following are likely sources of damage.

- Exposure to a deteriorating environment, such as cabin condensation, galley spillage, toilet spillage, cleaning fluids, and so forth.
- Contact between dissimilar metals
- Breakdown of surface protection

4.2.7 Accidental Damage

Accidental damage may be judged on the frequency of exposure and location of damage from such sources as:

- Ground-handling equipment
- Cargo-handling equipment
- Manufacturing deficiencies
- Improper maintenance and/or operating procedures
- Rain, hail
- Bird strike
- Runway debris
- Spillage
- Lightning strike

An example of the format used by AIRBUS for the A320 is available in the NARC for perusal. It shows how various ratings are collated to develop an inspection interval for a skin panel.

4.2.8 Systems and Power Plants

The logic for an MSI is divided into an analysis at the first level, where questions are asked as to whether the failure is evident, and then whether it is has a safety or economic effect. Answers to these questions follow the logic tree illustrated subsequently to place the failure into one of the following five categories:

1. Evident—Safety effect—tasks are required to assure safe operation
2. Evident—Operational economic effect
3. Evident—Nonoperational economic effect
4. Hidden—Safety effect
5. Hidden—Nonsafety economic effect.

The categorization is used to generate a scheduled task from one of the following:

- Lubrication/servicing—LU/SV
- Operational/visual check—OP/VC
- Inspection/functional check—IN/FC
- Restoration—RS
- Discard—DS

These are embraced in the maintenance activities considered in previous chapters, but also include provision for, and reliance on, crew and condition monitoring tasks. Tasks are generally considered in the aforementioned order, the top ones being easier and less cosily. Note that if a task is required and none of the aforementioned is effective, then redesign is mandatory. Unscheduled tasks may be generated also as outcomes of:

- The earlier scheduled tasks performed at their specified intervals
- Malfunction reports, from in the air or on the ground
- Data analysis, indicating adverse trends.

4.2.9 Setting Task Frequencies/Intervals

Sound information is necessary on which to base an effective interval for performing the aforementioned maintenance tasks. This information may come either from past experience on other similar systems, or from the manufacturer's test data. In the absence of any precedent, the working group is required to use its best judgement.

In pursuit of the aforementioned necessary data and experience, some items will be determined by the MRS to need a "threshold sample." In this case a specified number of items will be examined after a period in service to verify design calculations or empirical decisions made initially in respect of the maintenance tasks.

4.3 STATISTICAL RELIABILITY ASSESSMENT

Serviceability based on condition monitoring relies on information about likely or achieved reliability of the item being monitored. Failure rates are analyzed to establish the need for corrective action. Deterioration found during monitoring can also be used to prevent failure. However, failure must either be accepted as likely to occur and its consequences found either to be relatively unimportant, or safety assured by redundancy.

Failures may occur inflight, cause incidents, delays, engine shut downs, or air-turn-backs, and invoke the need to use redundant items. All these occurrences may be recorded to provide data. Information about an item can be derived from trends in the following types of data analysis:

- Flight hours/cycles (or MTBF)
- Dispatch reliability (or rate)
- Defect rate
- Removal rate or unscheduled removal rate
- Failure/fault rate
- Age bands at failure
- Probability of survival to a given life

Decisions regarding maintenance actions for items affected by changes in reliability may be made by people or committees associated with the reliability control process. In some airlines this function rests with the quality organization advised by engineering and maintenance sections. Procedures and lines of communication need to be well established and understood. Information may be presented in statistical process control formats where alert levels are set in advance. Adverse trends or penetration of safe levels become the trigger for remedial attention to the specified maintenance task.

4.4 LOGISTIC SUPPORT ANALYSIS

While the civil aviation maintenance analysis processes aforementioned were developed with an accent on commercial economic efficiency, military services were simultaneously setting out to improve operational availability with improved efficiency using much the same concepts. The F-111 System, first developed under then US Secretary of Defense Robert McNamara, was called the System Package Program, under which the aircraft and all its support requirements were to be provided by the prime contractor as a consistent whole. The US Air Force subsequently took a lead in the 1970s to establish Logistic Support Requirements on a contractual basis during development of new military systems. The consequent MIL-STDs 1388-2B represents the system the US government used for equipment development and acquisition projects.

Logistic support analysis (LSA) as a process can be applied very broadly to a wide range of support determinations and has standing beyond these military standards. A series of analytical LSA calculations can be made for four main purposes:

1. Set support criteria for system design
2. Allow evaluation of design options
3. Provisioning of support elements
4. Establish a baseline for predicting and measuring in-service support performance

While a purist's approach to LSA would concentrate on the analytical function, the Mil-Std tasks extend into determination of requirements for a more comprehensive integrated logistic support program; thus the following tasks include this form of output.

4.4.1 LSA Tasks

Supportability objectives to be achieved by a LSA program comprise 15 basic tasks in five groups. These are listed briefly as follows [4]:

1. Program planning and control
   1.1 Develop initial LSA strategy and define objectives
   1.2 Logistic support analysis plan (LSAP)
   1.3 Program reviews/design reviews
2. Mission and support systems definition
   2.1 Operational requirements and maintenance concept
   2.2 Standardization of equipment design constraints and criteria
   2.3 Comparative analysis of characteristics of equipment for supportability, readiness, and cost
   2.4 Technological opportunities for improved supportability
   2.5 Determine supportability-related design criteria (quantitative and qualitative)
3. Preparation and evaluation of alternatives
   3.1 Functional requirements identification (FMECA, RCM, etc.)
   3.2 Support system alternatives
   3.3 Evaluation of alternatives, trade-offs (level of repair analysis, or LORA)
4. Determination of logistic support resource requirements
   4.1 Operations and maintenance task analysis
   4.2 Initial support impact analysis
   4.3 Life-cycle support analysis postproduction
5. Supportability assessment
   5.1 Evaluation of performance, reporting, and validation

4.4.2 Failure Mode Effect Analysis

Failure mode effect and catastrophic analysis (FMECA) was developed in the 1950s and was one of the first systematic methods used to analyze failure in technical systems. The method has appeared under different names and with somewhat different content, such as FMEA. The difference between FMEA and FMECA is not distinct. Just as when you describe or rank criticality the various failures in failure modes and effect analysis, the analysis is often referred to as a FMECA. Criticality is a function of both the failure effect and frequency/probability.

In many countries, FMECA is nowadays developed as a national or military standard. In some enterprises, it has been a part of the design process, and the results from the analysis have been part of the system documentation. The purposes of performing FMEA are:

- Identification of single point failures and their effects on the process.
- Determining the magnitude of the effects of potential failures.
- Determining the criticality of each failure mode identified.

When performing FMEA, a checklist of failure modes (open, close, leaks, ruptures, etc.) is generated and applied to the various components of a system. The effect of the failure mode is determined by the system's response to the failure. The failure mode and effect are tabulated by consequences rarely examined. FMECA is effectively FMEA and risk ranking combined. For each failure mode identified, its likelihood and effects are assessed to determine its relative importance.

FMEA is a sample analysis method to reveal possible failure and predict failure effect on the system as a whole. The method is inductive; for each system component, we investigate what will happen if this component fails. The methods represent a systematic analysis of a system's components to identify all significant failure modes and to see how important they are, then assumed to function perfectly. FMEA is therefore suitable for revealing a critical combination of component failures.

FMEA is a "bottom-up" approach, particularly useful in examining the performance of relatively simple components, or for determining which types of failure are dangerous and which are safe, and finally for calculating overall failure rates to the two states for the complete component.

To ensure a systematic and full study of the system, a specific FMEA form is used. The content of the form may be suited to each application. The FMEA may, for example, include the columns shown in Table 4.1. The execution of an FMEA is illustrated by going through the form's content. The starting point for the analysis would be the information available from the system definition (functional diagrams, descriptions of components, etc.).

Identification (Column 1): Here the specific component is identified by a description and/or a number. It is also possible to refer to a system drawing or functional diagram.

Functional, operational state (Column 2): The function of the component, that is, its working tasks in the system, is briefly described. The state of the component when the system is in normal operation is described, for example, whether it is in continuous operation mode or in stand-by mode.

TABLE 4.1 FMECA Form

Analysis / Components	1 Identification	2. Function, operational state	3. Failure mode	4. Effect on other units in the system	5. Effect on the system	6. Corrective measures	7. Failure frequency	8. Failure effect ranking	9. Remarks
1									
2									
3									
…									

Failure modes (Column 3): All possible ways the component can fail to perform its function are listed in this column. Only the failure modes observable from "outside" are included. Internal failure modes are to be considered as causes of the failure. These possible causes are listed under a separate column. In some cases it will also be of interest to look at basic physical and chemical processes that can lead to failure (failure mechanisms), such as corrosion. Often we also state how different failure modes of the component are detected, and by whom.

For example, in a chemical process plant a specific valve is considered as a component in the system. The function of the valve is to open and close at demand. "The valve does not open on demand" and "the valve does not close on demand" are relevant failure modes, as well as "the valve opens when not intended" and "the valve closes when not intended." However, "washer bursts" is an example of the cause of a specific failure mode.

Effect on other units in the system (Column 4): Cases in which the specific failure mode affects other system components is stated in this column. Emphasis should be given to identification of failure propagation that does not follow the functional diagrams' functional chain. For example, increased load on remaining pillars supporting a common load when one pillar collapses; vibration in pump housing may induce failure of the pump drive unit, and so forth.

Effect on system (Column 5): In this column we describe how the system is influenced by the specific failure mode. The operational state of the system as a result of failure is to be stated, for example, whether the system is to remain in an operational state, change into another operational mode, or be in an unoperational state.

Corrective measures (Column 6): Here is described what has been done or what can be done to correct the failure, or possibly to reduce consequences of the failure. Measures may also be listed, aimed to reduce the probability that failure will occur.

Failure frequency (Column 7): In this column we state the estimated frequency (probability) for the specific failure mode and consequence. Instead of presenting frequencies for all failure modes, we may give a total frequency and relative frequency (in percentages) for different failure modes.

Failure effect ranking (Column 8): Here the failure is ranked according to its effect, with respect to reliability and safety, possibilities of mitigating the failure, length of repair time, and production loss, and so forth. We might, for example, use the following grouping of failure effects.

Small: A failure that does not reduce the system's functional ability beyond an acceptable level.
Significant: A failure that reduces the system's functional ability beyond an acceptable level, but with a consequence that can be corrected and controlled.
Critical: A failure that reduces the system's functional ability beyond an acceptable level and which creates an unacceptable condition, either operational or with respect to safety.
It is usually necessary to define the ranking in more detail in a specific situation.

Remarks (Column 9): Here we state, for example, any assumptions and suppositions.

By comparing failure frequency (probability) and failure effect (consequence), the criticality of a specific failure mode can be determined (Table 4.2).

TABLE 4.2 Criticality

	Consequence category		
Probability/frequency	**Small**	**Significant**	**Critical**
Very unlikely: Once per 100 year or more infrequently			
Unlikely: Once 100 year			
Quite likely: Once per 10 year			
Likely: Once per year			
Frequently: Once per month or more frequently			

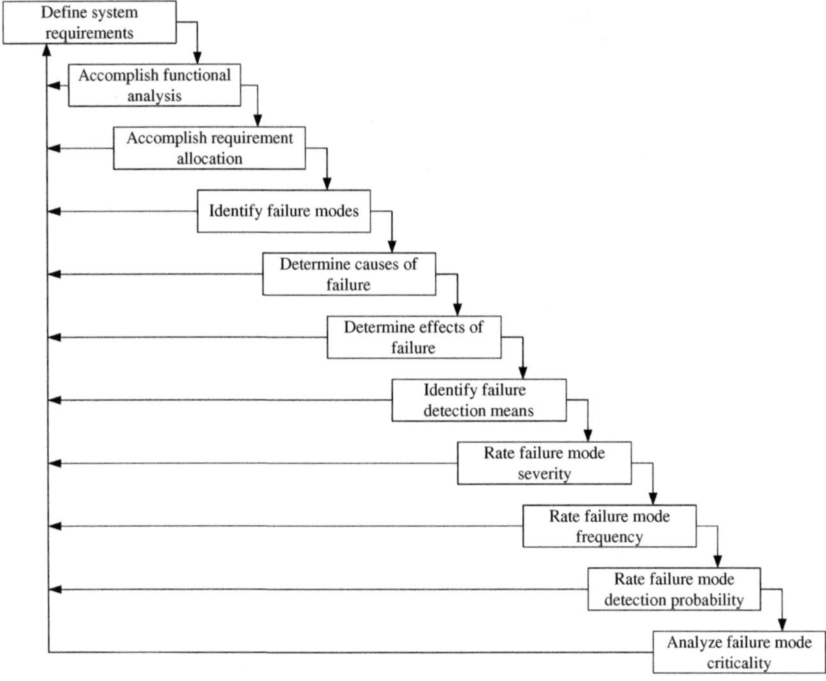

FIGURE 4.2 General approach to conducting a FMECA.

The procedure for executing a FMECA is shown in Fig. 4.2.

FMEA gives no guarantee that all critical failure modes have been revealed. There are many reasons for this, for example, lack of information and knowledge about the design and operation of components and the system. The analyst's imagination and ability to identify possible problems may of course also be a limiting factor. However, through the use of a systematic review as represented by FMECA, most system weaknesses resulting from individual component failure will be revealed. FMECA is the heart of the reliability-centered maintenance philosophy in that it identifies the consequences of failure leading to the same outcome in terms of maintenance requirements as the MSG-3 logic diagrams do, along very similar logic paths.

4.5 FAULT TREE ANALYSIS

The failure tree analysis (FTA) method was developed by Bell Telephone Laboratories in 1962 when they performed a safety evaluation of the Minuteman Launch Control System. The Boeing Company further developed the technique and made use of computer programs for both quantitative and qualitative fault tree analysis. Since the 1970s fault tree analysis has become very widespread and is today the most widely used reliability and risk analysis method.

Applications of the method now include most industries. The space industry and nuclear power industry have perhaps been the two where fault tree analysis is most widely used.

A fault tree analysis is a logical diagram, which shows the relation between system failures, that is, a specific undesirable event in the system, and failures of system components. The undesirable event constitutes the top event of the tree and different component failures constitute the basic events of the tree. For example, for a production process the top event might be that the process stops, and one basic event that one particular motor fails. A basic event does not necessarily represent a pure component failure; it may also represent human errors or failures due to external loads, such as extreme environmental conditions.

The results from the fault tree analysis include:

- A list of possible combinations of component failure/basic events that will ensure the top event occurs.
- Identification of critical component/events.
- The unreliability of the system, that is, the probability the top event will occur.

A fault tree analysis normally comprises several stages:

- Definition of the top event and framework conditions
- Construction of the fault tree
- Identification of the minimal cut sets
- Qualitative analysis of the fault tree
- Quantitative analysis of the fault tree

A fault tree comprises symbols that show the system's basic events, and the relation between these events and the state of the system. Graphic symbols showing the relationship are called logical gates. The output from a logical gate is determined by input state. The graphic symbols vary depending on the standard used. In the following we apply the American standard. Table 4.3 shows the most important symbols in the fault tree, with their interpretations.

A primary failure is often defined as a failure occurring under normal operation, a failure that the component itself is responsible for, while a secondary failure is defined as a failure not caused by the component itself, a failure occurring as a result of extreme environmental conditions, insufficient maintenance, and so forth, in which the causes of the failure are not studied further. The secondary failure is due to loading exceeding design specifications or because maintenance did not occur as prescribed.

Distinguishing between primary and secondary failure based on these definitions is often problematic and inappropriate. We will therefore in the following not distinguish between primary and secondary failure. A circle will be used to denote both failure types. Thus, a circle refers to a fault event sufficiently basic for no further development to be needed.

TABLE 4.3 Fault Tree Symbols

	Symbol	Interpretation
Logical symbols	"Or" gate	The output event A occurs if at least one of the input events E_i occurs. The number of input events is arbitrary.
	"And" gate	The output event A occurs if all input events E_i, occur. The number of input events is arbitrary.
	"Inhibit" gate	The output event A occurs if the input event E occurs and the condition B is present.
Basic (input) events	"Normal" input	Symbol for primary failure state.
	"Secondary" input	Symbol for secondary failure state.
Description of event	"Comment" rectangle	Events (states) are described in the rectangle. The rectangles are usually placed above all logical gates and input events.
Transfer symbols	"Transfer" in "Transfer"out	Transfer symbols for further development of a cause sequence; used when the same branch occurs at several places in the tree, and when the tree must be drawn on several pages.

The inhibit gate describes a casual relation between failure events. For the output event to occur, the input event must occur and a condition must be satisfied. An inhibit gate can be replaced by an "AND" gate.

A fault tree comprising of only "AND" and "OR" gates can be represented by a reliability block diagram. This is a logical diagram showing the functional ability of a system. Each component in the system is illustrated by a rectangle (Fig. 4.3).

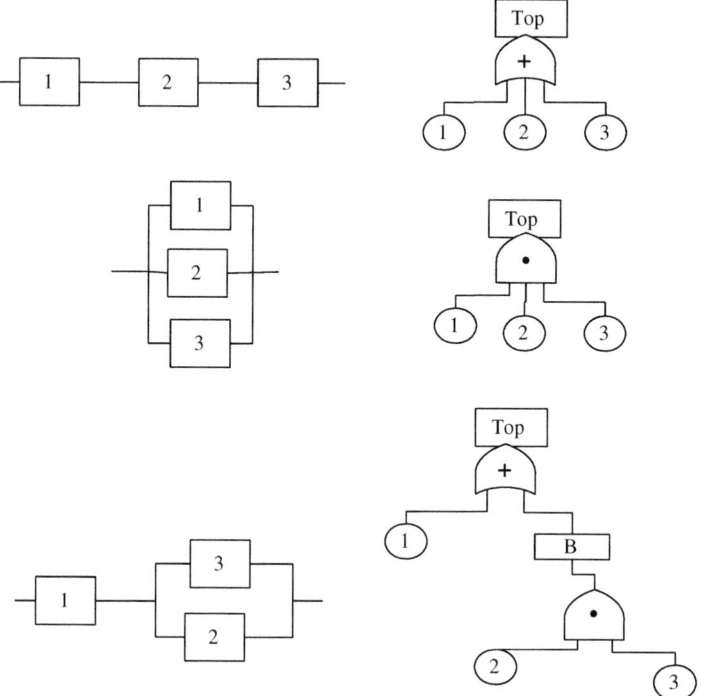

FIGURE 4.3 **Conversion between reliability block and fault tree.**

4.5.1 Qualitative Analysis of a Fault Tree

Once the fault tree is constructed, the structure of the tree can be examined qualitatively to understand the failure mechanism. This information is valuable as it provides a powerful insight into the identified possible modes of failure.

For a large fault tree, a more formal approach is needed. The principal means for fault tree evaluation is the complete set of minimal cut sets. System failure occurs when all the events in at least one minimal cut set occurs. The system can therefore be viewed as a series structure of the minimal cut parallel structure, as shown in Figs. 4.4 and 4.5.

4.5.2 Quantitative Analysis of a Fault Tree

If we can estimate the probability of the basic events of the fault tree, then we can perform a quantitative analysis. Usually we would like to calculate:

- The probability that the top event will occur
- The reliability importance (criticality) of the basic events (components) of the tree

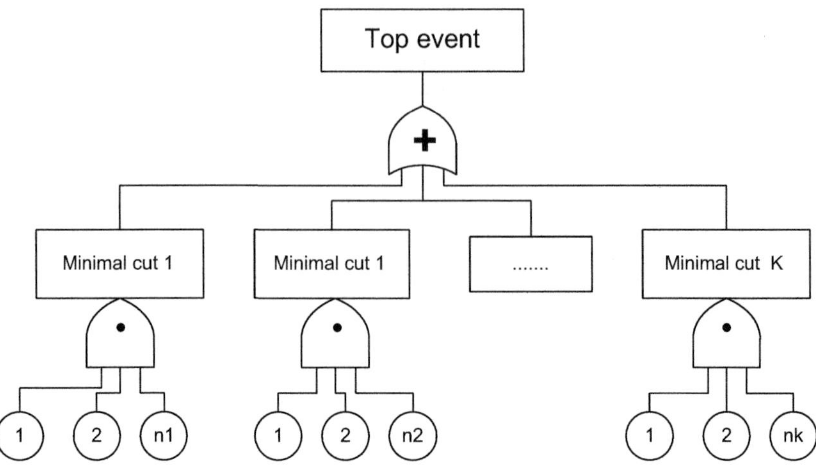

FIGURE 4.4 Minimal cut representation of the fault tree.

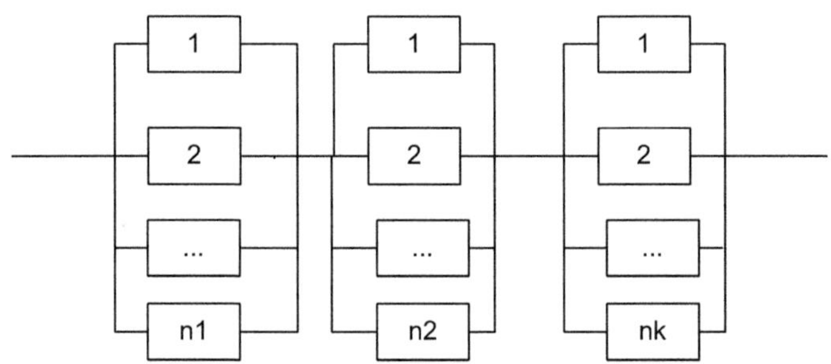

FIGURE 4.5 Minimal cut representation of the reliability block diagram.

A fault tree or a reliability block diagram provides a clear and well-arranged picture of the combination of equipment failures and other events leading to a specific undesirable system event. The fault tree is easily understood by persons with no knowledge of the technique. Fault tree analysis is well-documented and simple to use.

One of the advantages of using FTA is that the person undertaking the analysis is forced to understand the system. Many weak points in the system are revealed and corrected at the construction stage of the tree.

A fault tree gives a static "Picture" of the failure combination that can cause the top event to occur. However, the FTA method is unsuitable for analyzing a dynamic system. For example, it is difficult to analyze a standby system using fault tree analysis. Periodical testing and maintenance are also difficult to take into account in a fault tree analysis. Another problem is treatment of common-mode failure.

4.6 LEVEL OF REPAIR ANALYSIS

The policy for repair of an item may determine it to be unrepairable, partially repairable, or fully repairable. The process by which this is determined is called, in the US system, the level of repair analysis (LORA). The maintainability, costs, and support requirements, including skills, equipment, and so forth for each option for the item are determined and then an appropriate repair level policy decided. The analytical processes are complex, involving cost of repair and cost of replacement in the particular circumstances of the operator and will not be pursued in any depth.

4.7 LOGISTIC SUPPORT ANALYSIS RECORD

These tasks all feed data into the formal data-base known as the logistic support analysis record (LSAR). This is proscribed in most comprehensive detail in Mil-Std-1388-2B from which is derived a series of reports, which are listed as follows:

A Operation and maintenance requirements
B Item reliability and maintainability requirements
- Failure modes and effects analysis (FMECA)
- Criticality and maintainability analysis

C Operation and maintenance task summary
D Operation and maintenance task analysis
- Personnel and support requirements

E Support equipment and training equipment material description and justification
- Unit under test and automatic test program and training material description

F Facility description and justification
G Skill evaluation and justification
H Support items identification
I Transportability engineering characteristics

4.8 LSA MODELS

Blanchard [4] describes a number of logistic support analytical models evolved to undertake the analysis of most of the LSA processes. These are produced as commercial software products that can be sourced from various listed agencies in the United States. Most have code names, such as distributed integrated logistics support analysis (DILSA), which is a mini LSAR database processor, systems and logistics integration capability (SUC), and equipment designer's cost analysis system (EDCAS) for LORA. Practitioners in the use of LSA would undoubtedly find these models of assistance in structuring their task applications.

4.9 ELEMENTS OF ILS

The performance of a maintenance task that meets the standards required for airworthiness of an aircraft system requires support activities meeting the same exacting quality requirements. The elements of this support system were introduced in Sections 4.1, 4.2, and 4.5. Most aircraft operating organizations include a supply manager in their organization who is responsible for provisioning, procurement, warehousing, and transportation of support requirements. These functions contribute with engineering and maintenance to the broad functions of what is called integrated logistic support.

The main functions of an ILS system are:

- Configuration/data management
- Provisioning
- Procurement
- Inventory management
- Transportation
- Maintenance management
- Quality control

4.10 SUPPORT EQUIPMENT

The general category of support equipment may include a wide range of items, including precision electronic test equipment, mechanical test equipment, ground-handling equipment, special jigs and fixtures, maintenance stands, and so on. A feature of sound system design is the use, wherever possible, of existing or common types of versatile equipment that can then be used for a range of tasks.

Based on the maintenance plan for the aircraft system, the maintenance manager has to ensure availability of an adequate number of serviceable support equipment items for a wide range of tasks.

It is possible to use reliability data on a particular repairable item, plus the testing time or corrective maintenance time or each type of failure determined in the failure mode analysis to estimate the throughput of test or repair stations, and their utilization rate, and so estimate the number of these required. The serviceability and thus the availability of the test station itself can be a significant consideration in the availability of the aircraft systems it supports.

Calibration requirements for test instrumentation are an important factor. Many electronic instruments require periodic calibration against standards that ultimately need to be related and traceable to national standards. Mechanical and pressure measuring equipment can also need routine check calibration. The management of these processes is an additional load on maintenance control.

4.11 FACILITIES

Maintenance facilities can be critical to the effective maintenance support of an aircraft system. They can range from hangars and workshops for avionics equipment and engines to a paint shop and engine test facility or an aircraft wash-down site. They must be designed and equipped, if not already available, to provide adequate space and an environment suitable for the predicted task load. This assessment requires much of the maintenance data developed in the logistic support analysis leading to the maintenance plan. Specific requirements for types of power, environmental control for temperature, humidity and particle contamination, lighting, material handling, and personnel support needs all require consideration. The substantial costs and long lead time for approval and construction for facilities projects usually makes this aspect of maintenance planning one of the more urgent and difficult tasks. The task is not improved by lack of firm data in the early design stages of an aircraft project.

4.12 DATA

Prior to the advent of electronic data storage and retrieval systems the amount of paper specifications, drawing, manuals, and schedules required to support an aircraft type was a major problem; especially as much of it was being constantly revised. Access to some data is constrained by manufacturers' proprietary concerns and the initial planning for support must explore this area thoroughly. An established mode of communications with the holder of such data is a common way to overcome any availability problems. Satellite data transfer systems are useful for rapid responsiveness.

Many operators accept the cost of having manufacturers' field service representatives at their operating sites for the rapid resolution of information needs. Designers and manufacturers will normally maintain records of their equipment in service and be able to advise on fault patterns and defect resolution matters. A note of warning must be sounded over those occurrences that may lead to legal liability claims. Firm evidence is hard to obtain but it is generally accepted that in recognition of the litigious environment of aviation, some material that may be of support value to operators is not readily made available for this reason.

Chapter 5

Intelligent Structural Rating System Based on Back-Propagation Network

Chapter Outline

5.1 INTRODUCTION

Scheduled maintenance belongs to one of the major maintenance strategies: preventative maintenance, which is carried out at predetermined intervals to address any potential damage in case of failure. In the current civil aviation industry, scheduled maintenance programs are developed, mainly based on Maintenance Steering Group (MSG-3) logic and the initial maintenance intervals and tasks are specified in the Maintenance Review Board Report (MRBR), which outlines the minimum scheduled maintenance requirements for engines, systems, structures, and components of a given aircraft type in order to maintain their inherent economy, safety, and reliability [8]. Since most aircraft systems are provided by suppliers with their specific maintenance criteria, the determination of maintenance intervals for aircraft structures is considered one of the key tasks for the aircraft manufacturer at the design and manufacturing stage.

At every maintenance cycle, various inspection tasks are performed to detect damage and prevent structural degradation due to three damage sources throughout the operational life cycle, that is, accidental damage (AD), fatigue damage (FD), and environmental deterioration (ED). AD is described by the

Reliability Based Aircraft Maintenance Optimization and Applications
http://dx.doi.org/10.1016/B978-0-12-812668-4.00005-8

occurrence of a random discrete event that may undermine the inherent residual strength level. The high random characteristic of AD leads to great difficulty in assessing the susceptibility and detectability for SSIs. According to the requirement in the MSG-3 document, rating systems for AD should be established including the following evaluations:

1. Susceptibility to minor accidental damage based on frequency of exposure to and the location of damage from one or more sources.
2. Residual strength after accidental damage normally based on the likely size of damage relative to the critical damage size.
3. Timely detection of damage, based on the relative growth rate after damage is sustained and visibility of the SSI for inspection.

The AD sources are either internal or external and can be classified into two categories: manufacturing defects introduced during assembly and accidental damage introduced during operation and maintenance activities. For manufacturing defects, material properties need to be considered, and for accidental damage, structural maintainability, operational environment, and so forth should be taken into consideration. The selection of rating factors is based on four principles: operability, clarity, nonredundancy, and comparability [35]. Therefore, four main factors are considered as an example including: visibility, sensitivity to damage propagation, residual strength after damage, and likelihood of damage. Each main factor successively has several sub-factors, constructing the overall AD rating system as shown in Fig. 5.1.

The purpose of the rating system is to make proper AD inspection intervals. Generally, there are three methodologies including:

1. The matrix chart, which is developed from MSG-3 analysis and based on abundant practical engineering experience;
2. Modeling based on reliability data, which requires massive data collection on the similar type of aircraft;
3. Case-based reasoning (CBR), which is often applied to new aircraft. It requires tremendous data from maintenance cases; the more cases collected, the more accuracy CBR becomes.

Airbus Industries and the Boeing Company both apply the matrix chart to determine structural AD maintenance intervals. Factors' correlations are described by the form of a mathematical matrix, which can deal with multivariant problems [36]. A typical matrix chart is shown in Table 5.1. L and R are two groups of factors in alignment. The intersections of rows and columns represent the relationship of L_m and R_n. When the matrix chart is used to determine the inspection interval for aircraft structures, the factors may be varied due to different design concepts, structural diversity, and service environment.

The matrix chart used by Boeing Company and Airbus Industries is based on decades of structural experiments and in service experience. For new aircraft, such as the C919 under development, there is no original experience and

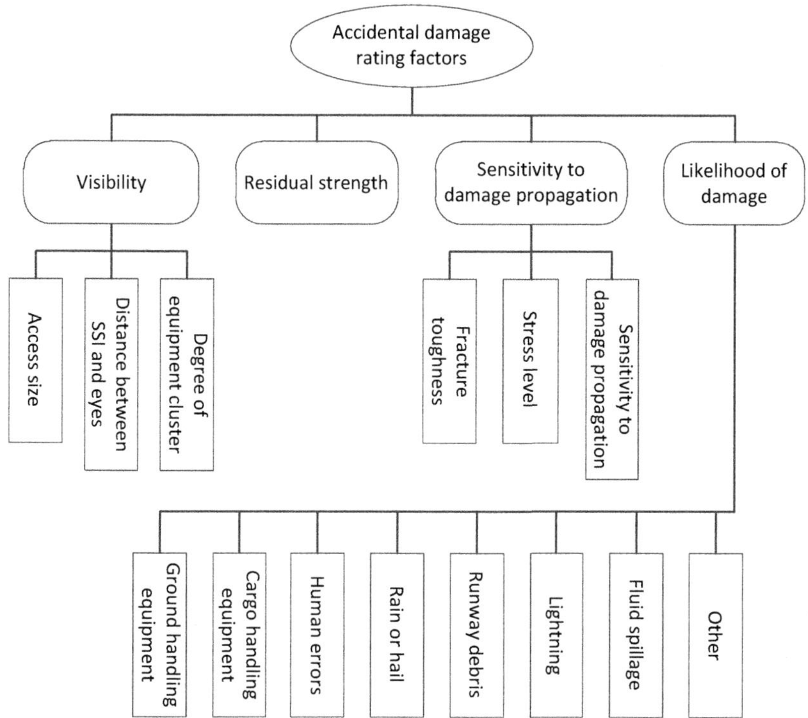

FIGURE 5.1 Accidental damage (AD) rating system.

TABLE 5.1 Rating Matrix Chart

		R			
		R_1	R_2	···	R_n
L	L_1				
	L_2		X_{22}		
	···				
	L_m				

a severe lack of information in operation and maintenance activities. Therefore, some other methodologies are proposed, such as modified analytical hierarchy process (AHP) based on rough sets [35] and CBR combined with fuzzy generalized nearest-neighbor matching [37]. A modified AHP was aimed at optimizing factors' weights, but it did not make much difference to the final results. The CBR was based on data from existing aircraft manufactures, which can be used only as a reference to make initial maintenance tasks.

This chapter introduces a new methodology to simulate the assumed AD rating system based on back-propagation network (BPN), which has the following advantages:

- The rating system can potentially have multiple data sources as input owing to its powerful data fusion capability.
- This method can also adjust the attributes' weights so that the factors importance can be reflected.
- This method can make predictions on inspection intervals based on similar cases and accumulated data from various sources due to its intelligent learning ability.

5.2 ARTIFICIAL NEURAL NETWORK

5.2.1 Basic Theory

Inspired by the information learning process in the human brain, the artificial neural network (ANN) is a kind of computer model to simulate the human pattern recognition function. An ANN is an interconnected group of nodes like the vast network of neurons in a brain. Basically, the dendrites of a biological neuron receive inputs from outside; main soma processes the inputs and then the axon outputs the result, as shown in Fig. 5.2.

ANNs are numerical structures consisting of massively parallel simple processing units widely linked with each other forming a network that can perform parallel processing and nonlinear transformation to model complex function relationship. They serve as alternative mathematical tools in many fields, such as system modeling, forecasting, pattern recognition, control and optimization, communication, and so forth [38]. A general perceptron model with weights and bias is shown in Fig. 5.3.

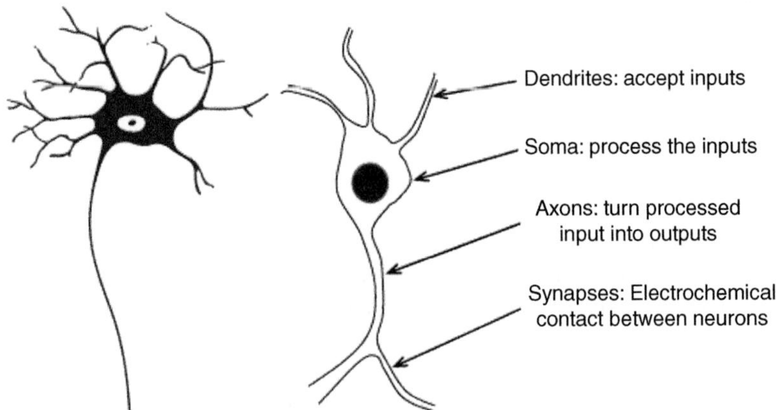

FIGURE 5.2 A biological neuron and main functions.

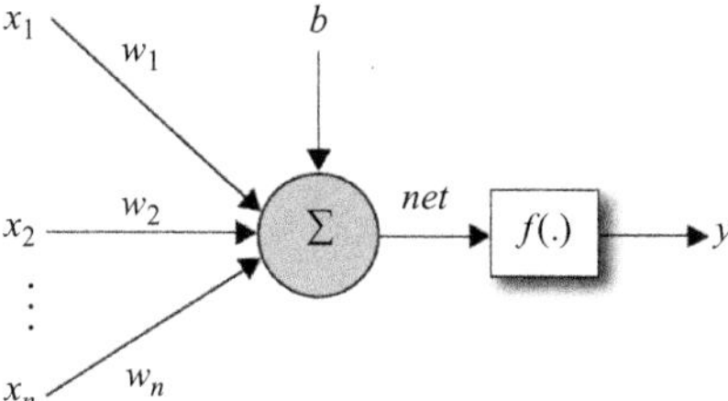

FIGURE 5.3 Perceptron model.

The term *weight* denotes the strength of the connection between two neurons, that is, the weight of information flowing from neuron to neuron. The first step is a process where the input $x_1, x_2, \ldots, x_n$ multiplied by their respective weights $w_1, w_2, \ldots w_n$ are summed by:

$$net = \left(\sum_{i=1}^{n} w_i \cdot x_i \right) \tag{5.1}$$

The *bias* is used to add or reduce the above summation value according to specific requirements expressed by the threshold value b. Then Eq. (5.1) is updated as:

$$net = \left(\sum_{i=1}^{n} w_i \cdot x_i \right) + b \tag{5.2}$$

A nonlinear activation function is usually included considering the variation of input conditions and their effects on the output so that the adequate amplifications can be used wherever necessary [39]. The final output of the neuron looks like:

$$y = f(net) \tag{5.3}$$

5.2.2 Back-Propagation Network

Among all kinds of ANN, the BPN is one of the most mature, widely used multilayer feed-forward neural networks based on error reverse spread. According to statistics, up to 80% of the neural network models apply BPNs or its variant forms, embodying the essence of neural networks [40]. A BPN includes at least an input layer, a hidden layer (implicit layer), and an output layer with a full connection between different layers but no links to neurons in the same layer. The input layer receives and distributes inputs. The hidden layer captures the nonlinear relationship of inputs and outputs. The output layer generates the calculated results. The BPN is developed on the basis of the back-propagation algorithm proposed in [41]. The network training is an unconstrained nonlinear

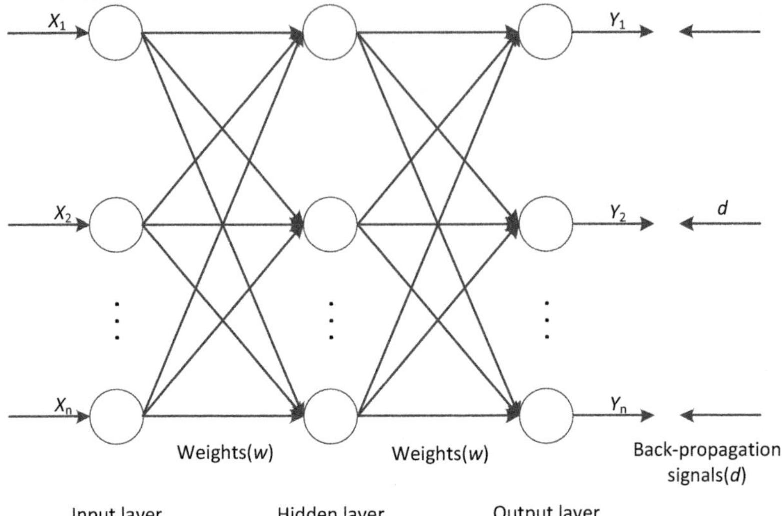

FIGURE 5.4 General structure of Back-propagation network (BPN).

minimization issue, and the goal of the training process is to adjust weights [38]. Theoretically, networks with biases, a sigmoid layer, and a linear output layer are capable of approximating any function with a finite number of discontinuities [42]. A hierarchical feed forward BPN frame is depicted in Fig. 5.4.

A single neuron can be described by this function:

$$Y_k = f\left(\sum_{i=1}^{m} w_{ik} x_i + b_k\right)$$

The BPN algorithm consists of two parts: information forward propagation and error back-propagation [43]. Assume p is the input with r neurons, $s1$ is the number of hidden neurons, $s2$ is the number of output neurons, and t is the target.

1. Information forward propagation.

The output of the ith neuron in the hidden layer:

$$a1_i = f1\left(\sum_{j=1}^{r} w1_{ji} p_j + b1_i\right), \quad i = 1,2,\ldots,s1$$

The output of the kth neuron in the output layer:

$$a2_k = f2\left(\sum_{i=1}^{s1} w2_{ik} a1_i + b2_k\right), \quad k = 1,2,\ldots,s2$$

Error function definition:

$$\xi = \frac{1}{2}\sum_{k=1}^{s2}(t_k - a2_k)^2, \quad e_k = t_k - a2_k$$

2. The weights adjust and error back-propagation is based on gradient descent.

Weights adjust in the output layer:

$$\Delta w2_{ik} = -\eta \frac{\partial \xi}{\partial w2_{ik}} = -\eta \frac{\partial \xi}{\partial a2_k} \cdot \frac{\partial a2_k}{\partial w2_{ik}} = \eta (t_k - a2_k) f2' a1_i = \eta \delta_{ik} a1_i$$

$$\delta_{ik} = (t_k - a2_k) f2' = e_k f2'$$

Biases adjust in the output layer:

$$\Delta b2_{ik} = -\eta \frac{\partial \xi}{\partial b2_{ik}} = -\eta \frac{\partial \xi}{\partial a2_k} \cdot \frac{\partial a2_k}{\partial b2_{ik}} = \eta (t_k - a2_k) f2' = \eta \delta_{ik}$$

Weights adjust in the hidden layer:

$$\Delta w1_{ji} = -\eta \frac{\partial \xi}{\partial w1_{ji}} = -\eta \frac{\partial \xi}{\partial a2_k} \cdot \frac{\partial a2_k}{\partial a1_i} \cdot \frac{\partial a1_i}{\partial w1_{ji}} = \eta \sum_{k=1}^{s2} (t_k - a2_k) f2' w2_{ik} f1' p_j$$
$$= \eta \delta_{ji} p_j$$

$$\delta_{ji} = e_i f1', \quad e_i = \sum_{k=1}^{s2} \delta_{ik} w2_{ik}$$

Biases adjust in the hidden layer:

$$\Delta b1_i = \eta \delta_{ji}$$

The algorithm is based on the minimization of errors, which is described as the difference between the expected output and actual result. The training process will finish when a certain accuracy level is met.

5.3 DESIGN BPN FOR AD

5.3.1 BPN Configuration

In AD rating analysis, previous solutions prefer to take it as a linear system, whereas in actuality it is a nonlinear mapping concept from influencing factors to a decision. The four integrated factors mentioned in the Section 5.1 are selected as input and the only output is the inspection interval. One hidden layer is defined with n neurons, as shown in Fig. 5.5. Generally speaking, the more

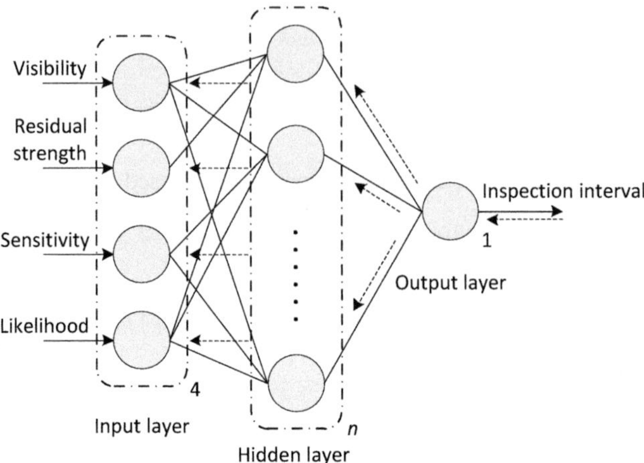

FIGURE 5.5 BPN for AD.

complicated the network is, the more hidden neurons are required, but currently there is no universal method. The selection principle of the number of nodes in the hidden layer will be addressed in the Section 5.4.

One of the advantages is that the BPN configuration fulfills the direct association between the rating factors and the final inspection interval. It saves a link in the matrix rating method that an intermediate rating score would be needed. This method considers the middle process as a black box, leaving it being processed automatically by machine learning. As a result, the developed AD rating system based on BPN has the potential of making best use of limited data from various sources with different formats and criteria.

5.3.2 Case Study

Twenty rating samples either based on engineering experience or taken from practical data are listed in Table 5.2.

In Table 5.2, R_V, R_S, R_{RS}, and R_L stand for visibility, sensitivity, residual strength, and likelihood, respectively. Numerical values 0, 0.5, 1, 1.5, 2 represent "low," "relatively low," "medium," "relatively high," and "high" for visibility and residual strength rating criterion, whereas these values represent "high," "relatively high," "medium," "relatively low," "low" for sensitivity and likelihood rating criterion. U denotes the number of samples. In the last column, for example, 1000FCs means the inspection interval is 1000 flight cycles.

First, the samples need to be pretreated because of the incompatible dimension. Herein, the sample data is normalized to fall in the interval [–1, 1] by using the following formula. The results are shown in Table 5.3.

$$\tilde{X}_{ij} = \frac{2\times[X_{ij} - min(X_{ij})]}{\left[max(X_{ij}) - min(X_{ij})\right]} - 1, \quad i = 1,2,3,\ldots,m, \quad j = 1,2,3,\ldots,n$$

TABLE 5.2 Visibility, Sensitivity, Residual Strength, and Likelihood Decision Rating

U	R_V	R_S	R_{RS}	R_L	I
1	0	1	1	0	2,000FCs
2	1	0	0	0	1,000FCs
3	1	1	1	0	4,000FCs
4	1	1	1	1	8,000FCs
5	2	1	2	1	16,000FCs
6	2	2	2	2	16,000FCs
7	2	1	1	1	16,000FCs
8	1	2	2	2	16,000FCs
9	0	2	1	0	4,000FCs
10	0	0	1	0	1,000FCs
11	0	1	1	1	4,000FCs
12	2	1	0	1	8,000FCs
13	1	2	1	0	8,000FCs
14	1	1	0	0	2,000FCs
15	1	2	1	2	16,000FCs
16	2	1	2	2	16,000FCs
17	1	0.5	1.5	1.5	8,000FCs
18	0	2	1	1	8,000FCs
19	1	0	2	2	16.000FCs
20	2	0	1	0	4,000FCs

where X, is the matrix of sample vectors including inputs and targets; and , the matrix of normalized sample vectors including inputs and targets.

Generally, there are three transfer functions: the hard-limit transfer function, the linear transfer function, and the sigmoid transfer function, which can be subdivided into the log-sigmoid function and tan-sigmoid function as shown in Fig. 5.6.

According to the normalization result, the tan-sigmoid function is selected between the input layer and the hidden layer. The linear function "purelin" is set between the hidden layer and the output layer without changing the magnitude of any exported value.

Set the number of hidden layer neurons to 10. The layer ratio of the network becomes 4:10:1. This network is trained by MATLAB software. The calculation parameters are set as follows:

- Maximum number of training epochs: 1000
- Learning rate: 0.01

TABLE 5.3 Normalization

U	R_V	R_S	R_{RS}	R_L	I
1	−1	0	0	−1	−0.8667
2	0	−1	−1	−1	−1
3	0	0	0	−1	−0.6
4	0	0	0	0	−0.0667
5	1	0	1	0	1
6	1	1	1	1	1
7	1	0	0	0	1
8	0	1	1	1	1
9	−1	1	0	−1	−0.6
10	−1	−1	0	−1	−1
11	−1	0	0	0	−0.6
12	1	0	−1	0	−0.0667
13	0	1	0	−1	−0.0667
14	0	0	−1	−1	−0.8667
15	0	1	0	1	1
16	1	0	1	1	1
17	0	−0.5	0.5	0.5	−0.0667
18	−1	1	0	0	−0.0667
19	0	−1	1	1	1
20	1	−1	0	−1	−0.6

- Performance goal: 10^{-2}
- Training function: trainscg

The problem of overfitting often occurs during the network training. Sometimes the error may go large when new data is applied in the situation that the error in the training set has already drew small. Therefore, the network fails to adapt to new data.

In MATLAB, the default method for improving generalization is "early stopping," which automatically divides the available data into three subsets as training, validation, and testing. Another advanced method is "regularization," which is suitable for cases with a small data set. Its essence is to modify the performance function by adding the mean of the sum of weights and biases to smooth the network response [42]. However, the outputs in this case study are a series of discrete values. Using "regularization" alone may result in a larger error for predicted results. The training network needs some criteria to stop the

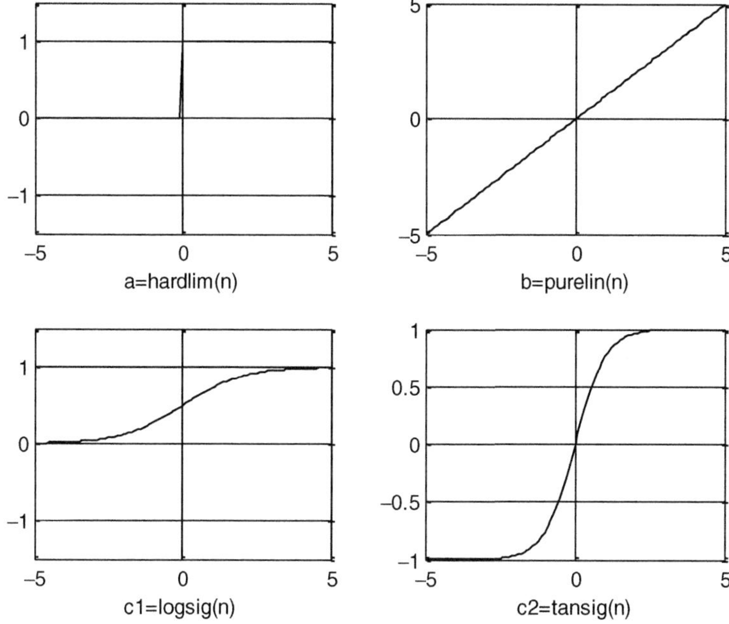

FIGURE 5.6 **Commonly used transfer functions of BPN.**

training procedure instead of being devoted to obtaining the minimum training error. Therefore, the regularization combined with early stopping is applied to improve the performance while controlling the overfitting problem in the validation process. The performance ratio in "regularization" is set to 0.6.

Three structural significant items (SSIs) on the wing (the upper rod, the safety pin, and the support bar connection) of a certain type of aircraft are selected to perform the prediction. The average values of inspection intervals for the three items are obtained by running the program 10 times. Details are listed in Table 5.4.

TABLE 5.4 Average Inspection Intervals and Final Approximation

SSI	Upper rod	Safety pin	Support bar connection
R_V	2	1	1
R_S	2	1.5	0.5
R_{RS}	1	1	1
R_L	2	1	1.5
P	16,372FCs	9,215FCs	8,503FCs
T	16,000FCs	8,000FCs	8,000FCs

TABLE 5.5 AD Rating according to [44]

SSI	Upper rod	Safety pin	Support bar connection
R_V	2	1	1
R_S	2	1	1
R_{RS}	1	1	1
R_L	2	2	2
ΣR	7	5	5
Inspection	16,000FCs	8,000FCs	8,000FCs

P indicates the average predicted inspection interval by BPN. It should be noticed that the preset inspection intervals are the following values: 1,000FCs, 2,000FCs, 4,000FCs, 8,000FCs, and 16,000FCs. Therefore, to make a proper approximation, conservative values are selected as the final results, denoted as *T*.

To demonstrate the effectiveness of the BPN method, the SSIs are rated according to the requirements in the Maintenance Program Development Policy and Procedures Handbook for Boeing 737NG, since these three structural items are similar cases. Results are presented in Table 5.5 AD rating according to [44].

5.4 DISCUSSION

The inspection intervals are a set of predetermined values, whereas the predicted results are a group of detailed data, which needs to be rounded to the nearest conservative values. It turns out that the final trimmed results exactly equal to the practical intervals for these SSIs, which demonstrates the applicability of the BPN on the AD rating system.

The error between the predicted interval and the preset value for safety pin is always larger than the upper rod and support bar connection. First, it is because of the sample set, in which some rating combinations correspond to the same interval. After training, there will be fluctuations within an acceptable range. Second, the reason is the design of the BPN. Trimming can be incorporated into the neural network training procedure to make more accurate predictions.

Usually samples are not used directly for network training due to different dimensions and singular values. A preprocessing, such as normalization or standardization of samples including input data and target data is often required. The objective of the preprocessing is to accelerate the convergence of the network training and make it more efficient. For the determination of the AD inspection interval, raw data can come from diverse sources, such as engineering experience, structural experiments, and in-service reliability data. The advantage of the BPN is that it provides a method to fuse raw data from different sources into one network through normalization and then perform specific training.

5.4.1 Selection of Number of Nodes in Hidden Layers and Parameter Ratio

The selection of number of neurons in hidden layers is a critical step that affects mapping capabilities. There is no unanimous method up until now. Several empirical formulas are often applied [38,40].

$$n = \sqrt{ml}$$
$$n = \sqrt{m + l} + a$$
$$n = \sqrt{0.43ml + 0.12l^2 + 2.54m + 0.77l + 0.35 + 0.51}$$

where n is the number of nodes in a hidden layer, m is the number of nodes in an input layer, l is the number of nodes in an output layer, and a is a constant within 1–10.

In terms of regularization, it is difficult to determine the optimum value for the performance ratio parameter r [42]. From the training window in MATLAB, the regression plot can be accessed, the analysis of which is a statistical process for estimating the relationship between the predicted value and the target value. The regression plot can be used to validate the network performance. The closer to 1 the regression value (R) is, the better the training results are indicated. Since the predicted results are all in an acceptable range, the maximization of the regression value of the test set is used as the optimization criterion; see Figs. 5.7 and 5.8

The surface in Fig. 5.7 represents the regression value R against the number of neurons n in the hidden layer ranging from 2 to 14 and the performance ratio r from 0.2 to 0.8. Each training experiment is trained at least 10 times to obtain the average R due to the variation characteristic introduced by the early

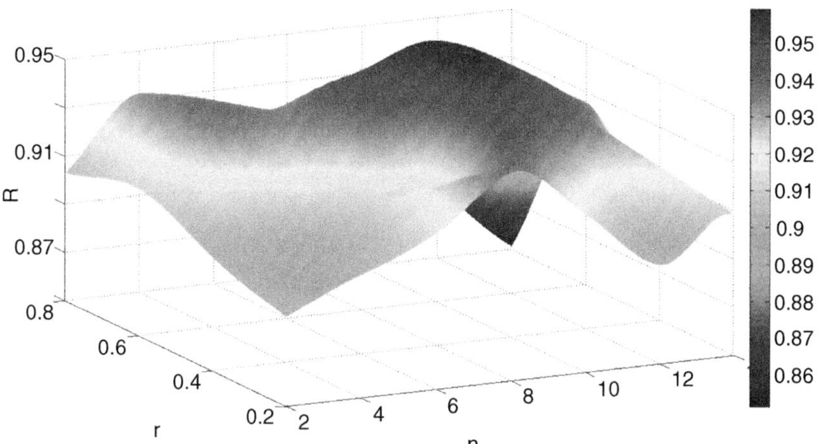

FIGURE 5.7 **Regression value R against hidden layer nodes n and performance ratio r.**

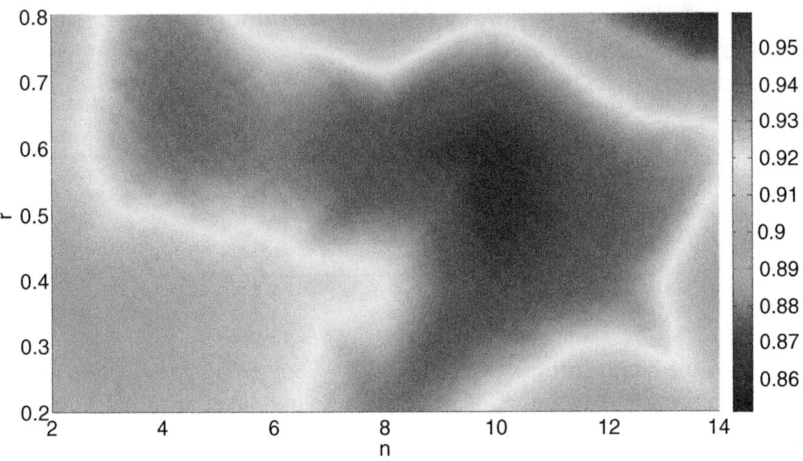

FIGURE 5.8 Contour plot of regression value *R*.

stopping. Fig. 5.8 is the contour plot of Fig. 5.7. It is shown that the fitting performs better when r is between 0.5 and 0.7, and n is around 8 to 12. The maximum R occurs when $n = 10$ and $r = 0.6$.

Therefore, the BPN for this AD rating system is optimized with three layers structure 4:10:1 and the performance ratio of 0.6.

5.4.2 Selection of Training Algorithms

The core of all the algorithms is the gradient of the performance function to determine how to adjust the weights to minimize performance error. There are several different training algorithms and their applicable problem types are roughly listed in Table 5.6.

In Table 5.6, BFGS represents Boryden, Fletcher, Goldfarb, and Shanno, the four researchers who contribute most to the quasi-Newton algorithm. In general, for function approximation problems and networks with fewer than 100 weights, the Levenberg–Marquardt algorithm (trainlm) has the fastest convergence but

TABLE 5.6 Training Algorithms for Two Problem Types

Pattern recognition	Function approximation
Resilient back-propagation	Levenberg–Marquardt
Scaled conjugate gradient	BFGS quasi-Newton
Conjugate gradient with Powell/Beale restarts	Scaled conjugate gradient
...	...

consumes a great amount of memory. As the number of weights increases, the advantage of trainlm decreases. The BFGS quasi-Newton method (trainbfg) also converges quickly and requires much memory, second to Levenberg-Marquardt. For recognition problems, the resilient back propagation (trainrp) is the fastest, and the memory requirement for this algorithm is relatively small compared to the other algorithms. The conjugate gradient algorithm, in particular scaled conjugate gradient (trainscg), performs well over a wide variety of problems, and consumes less memory than trainlm. The network in this study is used to train for function approximation with a small number of neurons. "Trainlm" and "trainbfg" converges too fast, thus "trainscg" is selected.

Last but not least, another hidden function of the BPN is to determine the attributes' weights. Scientifically speaking, the importance of the four integrated factors can be derivatively calculated by the BPN. Then the rating becomes a simple linear problem again after the nonlinear processing. The factors' weights were obtained by running the program many times in this case but it turned out that the values fluctuated due to nonlinearity. As continuing data accumulation, weights will be updated every now and then. The final objective is to predict the AD inspection intervals and the BPN initially fulfills the objective, leading to the unnecessary determination of the factors' weights.

5.5 CONCLUSIONS

This chapter proposed an artificial neural network to model the structural AD rating system, which is based on the assumption that certain data is available. A BPN for AD was established with four rating factors as inputs and the inspection interval as the output. The effectiveness of the new rating system was demonstrated in a case study, followed by a particular investigation into the training algorithm and parameters in order to achieve the best training result. As data accumulation from various sources becomes available and the rating times increase, the attributes' weights can be dynamically adjusted and therefore, the inspection interval can be updated to be more appropriate.

The BPN methodology developed in this chapter can be seen as an update of the structural rating system within MSG-3 analysis combining engineering experience and intelligent machine learning. The best advantage of the BPN for AD is the powerful, flexible data fusion and learning ability, which can be used to help make the MRBR for new aircraft structures when data is insufficient.

Chapter 6

Fault Tree Analysis for Composite Structural Damage

Chapter Outline

6.1 INTRODUCTION

In the past decade, the use of composite materials in commercial aircraft has grown significantly. More than 50% of the Boeing 787 and Airbus 350 airframes are made of composite materials [10,11]. The main motive is that composite is a lightweight material with design diversity. By selecting fiber material, fiber orientation, matrix volume, and so forth, the designer can manipulate the local material properties to increase the strength and resistance of the required direction [45]. However, such powerful design capabilities also present considerable side effects. Various combinations and forming processes induce high scatter in material properties and lead to complex damage modes, causing difficulties in fault diagnosis and prognosis.

Composite structures are usually fatigue and corrosion resistant but are more susceptible to impact damage caused by bird strike, hail and tools impact, and so forth. The fracture of composite structures is due to multiple damage modes and their interactions. The damage modes depend on various parameters, such as

Reliability Based Aircraft Maintenance Optimization and Applications
http://dx.doi.org/10.1016/B978-0-12-812668-4.00006-X

the property of the fiber and matrix, fiber lay-up, cure procedure, environment, temperature, operating conditions, and so forth. Due to a large scatter in material properties, deterministic methodologies may lead to conservative results, such as excessive weight and frequent inspections without taking account of uncertainties. Alternatively, probabilistic methodologies were proposed considering different aspects of the composite damages incorporating cumulative damage, manufacturing defects, operating environment and laminate theory, and so forth [46,47]. However, most of the studies are on a microscopic level based on experiments, computer modeling, or mechanical theory. Various macroscopic damages obtained from operational aircraft have not been comprehensively addressed. In this chapter, typical in-service damages occurring in composite airframes are collected via a survey to an airline maintenance department. It is noted that "damage mode" can have different meanings in different situations. Herein, "damage mode" refers to the superficial damage characteristics that can be seen visually or by nondestructive devices.

Traditionally, fault tree analysis (FTA) is a method used for system failures, which can dig out root causes and identify the weak links of a large system either qualitatively or quantitatively. This chapter extends FTA to areas of composite structures. A variety of damage modes and damage causes can be synthesized in a tree structure and analyzed systematically on a macroscopic level. Main damage modes and damage causes can be prioritized through qualitative analysis and, therefore, this method can be used as a diagnostic tool to identify and correct causes of composites failure. It can help promote understanding on complex damages and their logic relationship leading to failure more intuitively. Also, this method can be used for Monte Carlo simulation and fuzzy comprehensive evaluation if detailed damage information is available. Engineers from airlines and manufacturers can evaluate the reliability of the structure and the damage severity through extended quantitative analysis.

6.2 BASIC PRINCIPLES OF FAULT TREE ANALYSIS

FTA is one of the most important logic and probabilistic techniques used in probabilistic risk assessment and system reliability assessment. It was first developed by AT&T's Bell Laboratories in 1962. Later in 1974, US Atomic Energy Commission published a report on risk assessment of nuclear power stations, in which FTA was extensively and effectively used and the development of FTA was promoted greatly since then.

FTA is a deductive, "top-down" system evaluation process that focuses on one particular undesired event and possible causes through a qualitative model. The analysis starts with an undesired event with top-level hazard and identifies all credible single faults and faults combinations at the subsequent level that lead to the top event in a systematic pathway. Then the analysis continues through successive levels until a basic cause is unfolded or until the specific requirement is met. Basic cause events are such events that cannot be further broken down, which may be malfunctioning from the system inside or from external damage [48].

In other words, a fault tree is a graphic model of the pathways in a system leading to a foreseeable, undesirable fault event. Events and conditions that contribute to the undesirable event are interconnected through various logic symbols along the pathways to reflect their cause-and-effect relationship. This qualitative model is capable of conducting quantitative evaluations provided that numerical probabilities of occurrence are input and propagated throughout.

6.2.1 Elements of FTA

Basically, three kinds of event term are used in FTA:

1. *Basic event*: The initiating fault event without further development.
2. *Intermediate event*: A fault resulting from the logical interaction of initiating faults.
3. *Top event*: The occurrence of an undesired event for the system as a result of the occurrence of several intermediate events. Several combinations of initiating faults lead to the event.

A fault tree comprises two kinds of symbols: logic and event. The events are connected by various logic symbols representing different relationships. There is no connection within logic symbols or events. The general rule of symbols is to keep them simple and clear. Common fault tree symbols are listed in Table 6.1.

6.2.2 Boolean Algebra Theorems

Boolean algebra is used for set operation. Different from the common rule of operation, Boolean algebra can be used to analyze faults. In FTA, the occurrence of a top event can be described by combinations of occurrences of basic events. The minimal combination of basic events can be obtained through Boolean operation. Common Boolean operations are listed in Table 6.2.

6.3 FTA FOR COMPOSITE DAMAGE

Consider the damage of a composite structure as a system. The failure of the system is defined as one or more damages occurring in the structure leading to repair or replacement of the structure. The failure of the system is assumed to be the top event causing by both external and internal damage. "External damage" refers to any surface damage that is visible or barely visible, whereas "internal damage" denotes any damage that occurs inside the structure or throughout the structure that is either visible or detectable. External damage and internal damage can be subdivided into different damage modes as intermediate events. These intermediate events have various root causes as basic events. Two types of logic gates are used to connect different layers of the tree: the "AND" gate allows the output of

TABLE 6.1 Fault Tree Symbols

Symbol	Name	Definition
	Description box	Description of an output of a logic symbol or an event
	AND gate	Boolean Logic gate—event can occur when all the next lower conditions are true
	OR gate	Boolean Logic gate—event can occur if any one or more of the next lower conditions are true
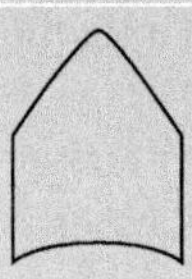	Priority AND gate	Boolean Logic gate—event can occur when all the next lower conditions occur in a specific sequence (sequence is usually represented by a conditional event)
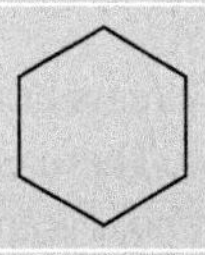	Inhibit	Output fault occurs if the (single) input fault occurs in the presence of an enabling conditional event
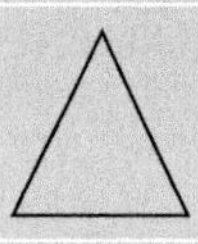	Transfer	Indicates transfer of information
	Basic event	Event which is internal to the system under analysis, requires no further development
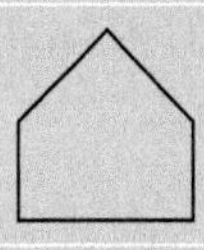	House	Event which is external to the system under analysis, it will or will not happen ($Pf = 1$ or $Pf = 0$)
	Conditional event	A condition that is necessary for a failure mode to occur

the event to occur only if all input events occur, which is equivalent to the Boolean symbol "·"; the "OR" gate allows the output of the event to occur if any one or more input events occur, which is equal to the Boolean symbol "+." A hierarchical fault tree can be established with proper gates connected. The advantage of this fault tree is that it provides an effective approach to synthesize various damage modes and damage causes in a systematic manner.

TABLE 6.2 Boolean Algebra Theorems

Name	Theorem description (X, Y, Z are sets)
Commutative law	$X \cdot Y = Y \cdot X, \quad X + Y = Y + X$
Associative law	$X \cdot (Y \cdot Z) = (X \cdot Y) \cdot Z, \quad X + (Y + Z) = (X + Y) + Z$
Distributive law	$X \cdot (Y + Z) = X \cdot Y + X \cdot Z, \quad X + (Y \cdot Z) = (X + Y) \cdot (X + Z)$
Absorption law	$X \cdot (X + Y) = X, \quad X + (X \cdot Y) = X$
Complementation law	$X + \overline{X} = U, \quad X \cdot \overline{X} = \Phi, \quad \overline{\overline{X}} = X$
Idempotency law	$X \cdot X = X, \quad X + X = X$
De Morgan's law	$\overline{(X \cdot Y)} = \overline{X} + \overline{Y}, \quad \overline{(X + Y)} = \overline{X} \cdot \overline{Y}$

A survey was conducted at an airline maintenance department to collect information on in-service damage in composite airframes. A typical composite laminated panel made of carbon fiber reinforced plastic (CFRP) was selected as an illustration. The overall organization of the fault tree is shown in Fig. 6.1.

The top event is the failure of the CFRP laminate panel, followed by external and internal damage connected by an AND gate as the first layer of intermediate events. The intermediate events on the second layer are various damage modes connected by two OR gates with the upper layer. The basic events are all

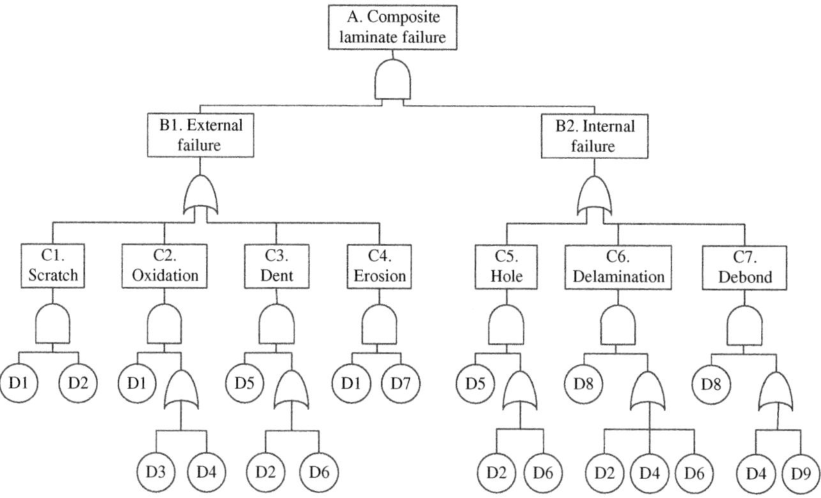

FIGURE 6.1 Fault tree construction of composite laminate structure. *D1*, Surface protection; *D2*, mishandling; *D3*, lightning strike; *D4*, heat; *D5*, material resistance; *D6*, natural object impact; *D7*, wind/sand/rain erosion; *D8*, manufacturing defects; *D9*, overloading.

potential damage sources or root causes. Take C3 Dent as an illustration: it is caused by both D5 Material resistance and the other event, which is caused by either D2 Mishandling or D6 Natural object impact. According to the survey, C1 to C7 are seven of the most frequent damage modes that occurred in aircraft composite structures made of CFRP. Crack is not listed because the occurrence of several other damage modes, such as dent, hole, and delamination are accompanied by fiber buckling, matrix cracks, and even fiber breakage [49]. It should be mentioned that moisture and ultraviolet radiation is not included because carbon fiber reinforced plastics have low sensitivity to the environment [8]. If the selected composite panel is a honeycomb or is made of Kevlar, moisture ingress and ultraviolet radiation will be significant contributors to the damage. Most of the delamination and debonding is not only due to impact damage, heat, and overloading, but also caused by defects during manufacturing [50]. Therefore, manufacturing defect is considered as an important contributor. Impact damage, such as dent covers a wide variety of events including tool drop, cargo buggy strike, bird strike, and so forth. For the sake of simplicity, impact damage sources are divided into two categories: human errors and natural accidents. Meantime, other damage sources have been simplified to facilitate the qualitative analysis.

In addition to the damage sources, the properties of the composite material and surface protection are also taken into account as basic events. Material resistance is one of the inherent properties of composite laminates, which is the ability of the material to resist impact damage [51]. As to surface protection, abrasion resistant coatings, antierosion coatings, antistatic coatings, and so forth can effectively reduce the damage caused by scratching, lightning strike, and so forth.

6.4 QUALITATIVE ANALYSIS

The primary step of the qualitative process is to obtain a minimal cut set list, which provides key qualitative information. Three importance analyses including structure importance analysis, probability importance analysis, and relative probability importance analysis are performed sequentially on the basis of the minimal cut sets.

6.4.1 Minimal Cut Sets

The minimal cut sets for the top event are a group of sets consisting of the smallest combinations of basic events that result in the occurrence of the top event. They represent all the ways in which the basic events cause the top event [52]. The equivalent Boolean algebra function of Fig. 6.1 can be expressed as:

$$
\begin{aligned}
A &= \mathrm{B1} \times \mathrm{B2} = (\mathrm{C1+C2+C3+C4}) \times (\mathrm{C5+C6+C7}) \\
&= [\mathrm{D1D2+D1(D3+D4)+D5(D2+D6)+D1D7}] \\
&\quad \times [\mathrm{D5(D2+D6)+D8(D2+D4+D6)+D8(D4+D9)}]
\end{aligned} \tag{6.1}
$$

By applying the equivalent Boolean algebra operation, the final Boolean expression of the top event can be obtained as:

$$\begin{aligned} A &= (\text{D2D5} + \text{D5D6}) + (\text{D1D2D8} + \text{D1D4D8}) \\ &+ (\text{D1D3D6D8} + \text{D1D3D8D9} + \text{D1D6D7D8} + \text{D1D7D8D9}) \end{aligned} \tag{6.2}$$

It can be seen from Eq. (6.2) that the top event is composed of two second-order minimal cut sets: $K_1 = \{\text{D2, D5}\}$, $K_2 = \{\text{D5, D6}\}$; two third-order minimal cut sets: $K_3 = \{\text{D1, D2, D8}\}$, $K_4 = \{\text{D1, D4, D8}\}$; and four fourth-order minimal cut sets: $K_5 = \{\text{D1, D3, D6, D8}\}$, $K_6 = \{\text{D1, D3, D8, D9}\}$, $K_7 = \{\text{D1, D6, D7, D8}\}$, $K_8 = \{\text{D1, D7, D8, D9}\}$. All the eight minimal cut sets are the premise of the following three importance analyses.

6.4.2 Structure Importance Analysis

Structure importance analysis is used to analyze the degree of importance of every basic event influencing the top event, from the perspective of the fault tree structure itself, regardless of the probability of the basic event [53]. There are two ways to perform the analysis. One is to calculate the structure importance coefficient for every basic event. The other is to estimate the importance by minimal cut sets. The complexity of the first method is increased by the growing number of basic events, in this case 29 combinations. Therefore, the second method by minimal cut sets is applied. The importance coefficient of the basic event X_i is estimated by:

$$I_{(i)} = \sum_{X_i \in K_j} \frac{1}{2^{n_i - 1}} \tag{6.3}$$

Where $I_{(i)}$ is the estimation value of the structure importance of the basic event X_i; $X_i \in K_j$ is the basic event X_i, which belongs to minimal cut set K_j; and n_i is the number of events in the minimal cut set containing X_i. Take D6, for example, the minimal cut sets containing D6 are K_2, K_5, and K_7. The number of events in each set is 2, 4, and 4, respectively. Thus, the structure importance coefficient $I_{(6)} = \frac{1}{2^{2-1}} + \frac{1}{2^{4-1}} + \frac{1}{2^{4-1}} = \frac{3}{4}$. After calculation, the results are shown in Table 6.3.

This table illustrates that surface protection (D1), material resistance (D5), and manufacturing defects (D8) play the most important roles. In terms of the

TABLE 6.3 Results of Structure Importance Analysis

1	Surface protection (D1)	Material resistance (D5)	Manufacturing defects (D8)	
3/4	Mishandling (D2)		Natural object impact (D6)	
1/4	Lightning (D3)	Heat (D4)	Wind erosion (D7)	Overloading (D9)

accidental damage sources, impact damage caused by mishandling and natural object impact is the main cause. Other damage sources are relatively less important.

6.4.3 Probability Importance Analysis

Probability importance is the derivative of the probability of the top event to the basic event, thereby reflecting the influence of the unreliability of the basic event to that of the top event. If the probability of the top event is $P(A) = Q(p_1, p_2, \cdots p_n)$, $n \in N^+$, the probability importance of the basic event D_i is expressed as:

$$I_p(\mathrm{D}_i) = \frac{\partial Q(p_1, \cdots p_n)}{\partial p_i} \quad i = 1, \cdots, n \tag{6.4}$$

Let $p(X_i)$ denote the probability of the basic event X_i, then the probability of the top event A is calculated as:

$$\begin{aligned} P(A) = & \sum_{i=1}^{8} p(K_i) - \sum_{i<j=2}^{8} p(K_i K_j) + \sum_{i<j<k=3}^{8} p(K_i K_j K_k) - \sum_{i<j<k<l=4}^{8} p(K_i K_j K_k K_l) \\ & + \sum_{i<j<k<l<m=5}^{8} p(K_i K_j K_k K_l K_m) - \sum_{i<j<k<l<m<n=6}^{8} p(K_i K_j K_k K_l K_m K_n) \\ & + \sum_{i<j<k<l<m<n<o=7}^{8} p(K_i K_j K_k K_l K_m K_n K_o) + p(K_1 K_2 K_3 K_4 K_5 K_6 K_7 K_8) \end{aligned} \tag{6.5}$$

where $p(K_i)$ can be obtained by Eq. (6.6):

$$p(K_i) = \prod_{i \in K_i} p(X_i) \tag{6.6}$$

where K_i is the ith minimal cut set, $i = 1, 2, \ldots, 8$.

According to the rare event approximation [54], $P(A)$ can be approximated to its first item $\sum_{i=1}^{8} p(K_i)$. Therefore, the probability importance of each basic event is calculated as:

$$\begin{aligned} I_p(\mathrm{D1}) = & \, p(\mathrm{D2})p(\mathrm{D8}) + p(\mathrm{D4})p(\mathrm{D8}) + p(\mathrm{D3})p(\mathrm{D6})p(\mathrm{D8}) \\ & + p(\mathrm{D3})(\mathrm{D8})(\mathrm{D9}) + p(\mathrm{D6})p(\mathrm{D7})p(\mathrm{D8}) + p(\mathrm{D7})p(\mathrm{D8})p(\mathrm{D9}) \end{aligned}$$

$$I_p(\mathrm{D2}) = p(\mathrm{D5}) + p(\mathrm{D1})p(\mathrm{D8})$$

$$I_p(\mathrm{D3}) = p(\mathrm{D1})p(\mathrm{D6})p(\mathrm{D8}) + p(\mathrm{D1})p(\mathrm{D8})p(\mathrm{D9})$$

$$I_p(\mathrm{D4}) = p(\mathrm{D1})p(\mathrm{D8})$$

$$I_p(\mathrm{D5}) = p(\mathrm{D2}) + p(\mathrm{D6})$$

$$I_p(\mathrm{D6}) = p(\mathrm{D5}) + p(\mathrm{D1})p(\mathrm{D3})p(\mathrm{D8}) + p(\mathrm{D1})p(\mathrm{D7})p(\mathrm{D8})$$

$$I_p(\mathrm{D7}) = p(\mathrm{D1})p(\mathrm{D6})p(\mathrm{D8}) + p(\mathrm{D1})p(\mathrm{D8})p(\mathrm{D9})$$

$$I_p(\mathrm{D8}) = p(\mathrm{D1})p(\mathrm{D2}) + p(\mathrm{D1})p(\mathrm{D4}) + p(\mathrm{D1})p(\mathrm{D3})p(\mathrm{D6}) \\ + p(\mathrm{D1})p(\mathrm{D3})p(\mathrm{D9}) + p(\mathrm{D1})p(\mathrm{D6})p(\mathrm{D7}) + p(\mathrm{D1})p(\mathrm{D7})p(\mathrm{D9})$$

$$I_p(\mathrm{D9}) = p(\mathrm{D1})p(\mathrm{D3})p(\mathrm{D8}) + p(\mathrm{D1})p(\mathrm{D7})p(\mathrm{D8})$$

Except for D1 Surface protection, D5 Resistance, and D8 Manufacturing defects, all the other basic events are in practice small probability events. Thus, it is relatively easy to make qualitative comparisons.

Since

$$I_p(\mathrm{D3}) = I_p(\mathrm{D7}) = p(\mathrm{D1})p(\mathrm{D8})[p(\mathrm{D6}) + p(\mathrm{D9})]$$

According to the associative law of addition,

$$I_p(\mathrm{D9}) = p(\mathrm{D1})p(\mathrm{D8})[p(\mathrm{D3}) + p(\mathrm{D7})]$$

Generally, D6 Natural object impact is one of the main damage sources with a frequency that is much higher than D3 Lightning, D7 Erosion, and D9 Overloading. So,

$$p(\mathrm{D6}) + p(\mathrm{D9}) > p(\mathrm{D3}) + p(\mathrm{D7})$$

Thus,

$$I_p(\mathrm{D3}) = I_p(\mathrm{D7}) > I_p(\mathrm{D9})$$

Because of the small probability principle,

$$p(\mathrm{D5}) + p(\mathrm{D1})p(\mathrm{D8}) > p(\mathrm{D5}) + p(\mathrm{D1})p(\mathrm{D8})[p(\mathrm{D3}) + p(\mathrm{D7})]$$

$$p(\mathrm{D1})p(\mathrm{D8}) > p(\mathrm{D1})p(\mathrm{D8})[p(\mathrm{D6}) + p(\mathrm{D9})]$$

Therefore,

$$I_p(\mathrm{D2}) > I_p(\mathrm{D6})$$

$$I_p(\mathrm{D4}) > I_p(\mathrm{D3})$$

Since D5 Material resistance is one of the inherent properties of the composite structure, which is difficult to change. $p(\mathrm{D5})$ is considered as a large probability. Then

$$p(\mathrm{D5}) + p(\mathrm{D1})p(\mathrm{D8})[p(\mathrm{D3}) + p(\mathrm{D7})] > p(\mathrm{D1})p(\mathrm{D8})$$

TABLE 6.4 Results of Probability Importance Analysis

High	Mishandling (D2)	Natural object impact (D6)	
Medium	Lightning (D3)	Heat (D4)	Erosion (D7)
Low	Overloading (D9)		

So,

$$I_P(\text{D6}) > I_P(\text{D4})$$

The final inequality and the results are obtained and shown in Table 6.4.

$$I_p(\text{D2}) > I_p(\text{D6}) > I_p(\text{D4}) > I_p(\text{D3}) = I_p(\text{D7}) > I_p(\text{D9}) \tag{6.7}$$

This table suggests that D2 Mishandling and D6 Natural object impact are the most critical damage sources, followed by D4 Heat, D3 Lightning, and D7 Erosion. D9 Overloading ranks last. According to the survey, most mishandlings lead to either apparent damage, such as scratch, dent, or internal damage, such as delamination. Natural object impact, such as runway debris is less likely to happen compared to human error. These two damage categories by human behavior and natural accidents are the most severe damages, which are of particular concern.

It should be noted that since D1 Surface protection, D5 Resistance, and D8 Manufacturing defects are inherently related to material properties or manufacturing process. It is difficult to define their probabilities, which will be discussed separately.

6.4.4 Relative Probability Importance Analysis

Probability importance analysis determines the influence of the probability change of the basic event on that of the top event, but cannot represent the difficulty of different basic events' improvement. Relative probability importance analysis is introduced to measure the variation of the top event probability from the aspects of sensitivity and probability of the basic event itself [55,56].

$$I_c(D_i) = \frac{p_i}{Q(p_1,\cdots p_n)} \cdot \frac{\partial Q(p_1,\cdots p_n)}{\partial p_i} \quad i = 1,\cdots,n \tag{6.8}$$

From Eq. (6.8), the relative probability importance of each basic event is calculated as follows:

$$\begin{aligned} I_c(\text{D1}) = [&p(\text{D1})p(\text{D2})p(\text{D8}) + p(\text{D1})p(\text{D4})p(\text{D8}) + p(\text{D1})p(\text{D3})p(\text{D6})p(\text{D8}) \\ &+ p(\text{D1})p(\text{D3})(\text{D8})(\text{D9}) + p(\text{D1})p(\text{D6})p(\text{D7})p(\text{D8}) \\ &+ p(\text{D1})p(\text{D7})p(\text{D8})p(\text{D9})]/Q(p_1,\cdots,p_n) \end{aligned}$$

$$I_c(\text{D2}) = [p(\text{D2})p(\text{D5}) + p(\text{D1})p(\text{D2})p(\text{D8})]/Q(p_1,\cdots,p_n)$$

$$I_c(\text{D3}) = [p(\text{D1})p(\text{D3})p(\text{D6})p(\text{D8}) + p(\text{D1})p(\text{D3})p(\text{D8})p(\text{D9})]/Q(p_1,\cdots,p_n)$$

$$I_c(\text{D4}) = p(\text{D1})p(\text{D4})p(\text{D8})/Q(p_1,\cdots,p_n)$$

$$I_c(\text{D5}) = [p(\text{D2})p(\text{D5}) + p(\text{D5})p(\text{D6})]/Q(p_1,\cdots,p_n)$$

$$I_c(\text{D6}) = [p(\text{D5})p(\text{D6}) + p(\text{D1})p(\text{D3})p(\text{D6})p(\text{D8}) + p(\text{D1})p(\text{D6})p(\text{D7})p(\text{D8})]/Q(p_1,\cdots,p_n)$$

$$I_c(\text{D7}) = [p(\text{D1})p(\text{D6})p(\text{D7})p(\text{D8}) + p(\text{D1})p(\text{D7})p(\text{D8})p(\text{D9})]/Q(p_1,\cdots,p_n)$$

$$I_c(\text{D8}) = [p(\text{D1})p(\text{D2})p(\text{D8}) + p(\text{D1})p(\text{D4})p(\text{D8}) + p(\text{D1})p(\text{D3})p(\text{D6})p(\text{D8}) + p(\text{D1})p(\text{D3})p(\text{D8})p(\text{D9}) + p(\text{D1})p(\text{D6})p(\text{D7})p(\text{D8}) + p(\text{D1})p(\text{D7})p(\text{D8})p(\text{D9})]/Q(p_1,\cdots,p_n)$$

$$I_c(\text{D9}) = [p(\text{D1})p(\text{D3})p(\text{D8})p(\text{D9}) + p(\text{D1})p(\text{D7})p(\text{D8})p(\text{D9})]/Q(p_1,\cdots,p_n)$$

Similar comparisons can be made and it is shown that D2 Mishandling and D6 Natural object impact rank the highest irrespective of the particular group of events mentioned previously (D1, D5, and D8).

6.5 QUANTITATIVE ANALYSIS

The survey collected damage records on aircraft structures made of composite materials. Wing structural damage (ATA Chapter 57) of two types of aircraft fleet (Boeing 737-800 and Boeing 757-200) recorded over a 10-year period were obtained. A breakdown of damage categories and their numbers of occurrence on composites made with CFRP are plotted in Fig. 6.2. It is shown that dent is the most frequent damage mode followed by painting peel-off. Due to the inconsistency of maintenance recording, damages, such as dent, scratch, erosion, and so forth can all lead to painting peel-off. To facilitate the following analysis, painting peel-off caused by scratch is assumed to take up approximately half of the percentage, rounding to 12%.

A statistical analysis was performed aiming at the selected laminated CFRP panel. Twelve occurrences of the primary damage mode dent were recorded in the CFRP panel in 6 Boeing 737-800 aircraft. The design life of Boeing 737-800 is 100,000 flight hours and the composite panel is assumed to have the same design life as the airplane. Therefore, the average number of dent events per flight hour is 2e-5. According to the percentage distribution of each damage mode in Fig. 6.2, the average numbers of occurrence for the seven damage modes per flight hour were calculated with some rounding and are shown in Table 6.5.

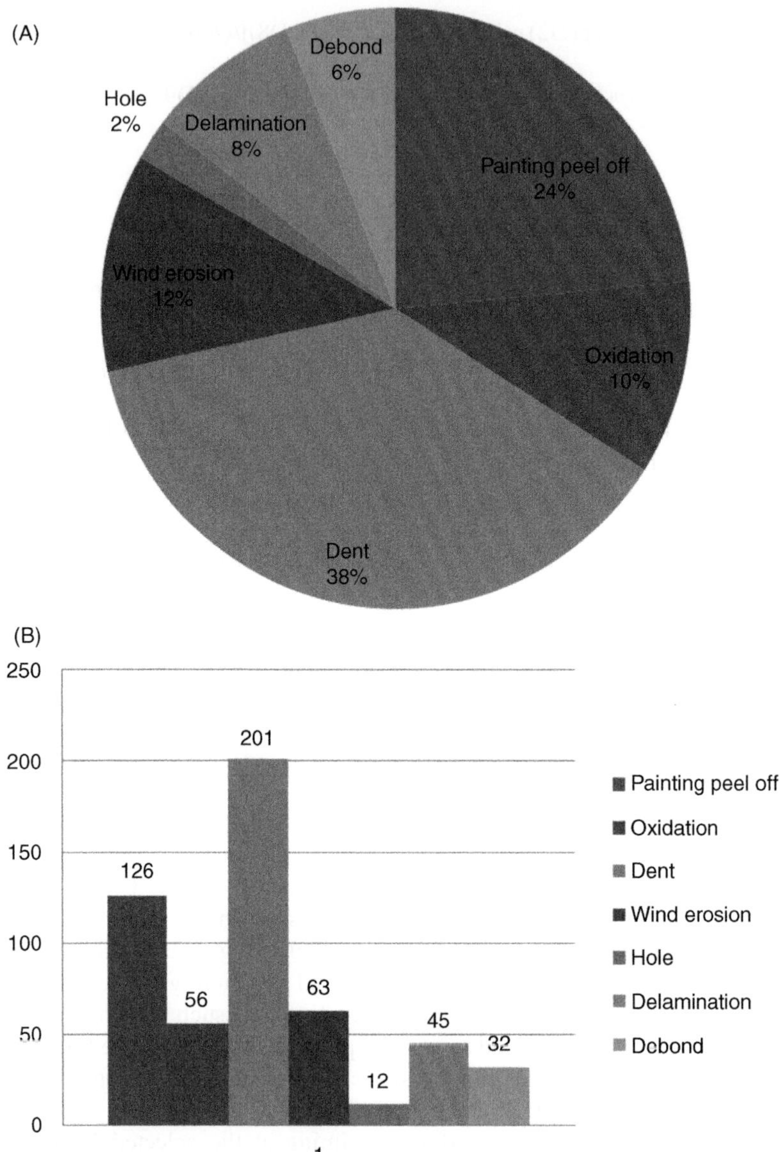

FIGURE 6.2 (A) Damage category and (B) occurrence number of CFRP composites.

Further, the probability of occurrence for damage modes was distributed to various damage causes by engineering experience from the airline. The distribution law is based on the Boolean operation in the fault tree structure in Fig. 6.1. Take C3 Dent as an illustration: it is caused by both D5 Material resistance and the other intermediate event, which is caused by either D2 Mishandling or D6 Natural object impact. Then we have the following relationship:

TABLE 6.5 Probability of Occurrence for Damage Modes

Damage mode	Number of occurrence per flight hour (probability)
C1 scratch	0.6e-5
C2 oxidation	0.5e-5
C3 dent	2e-5
C4 erosion	0.6e-5
C5 hole	0.1e-5
C6 delamination	0.4e-5
C7 debond	0.2e-5

$$P(C3) = P(\mathrm{D5}) \times [P(\mathrm{D2}) + P(\mathrm{D6})] \quad (6.9)$$

Replacing by numerical values, the allocated probability of occurrence for every basic event is obtained in the fault tree as listed in Table 6.6.

Once the probability distributions were assigned to every basic event, Monte Carlo simulation was then conducted as a validation of the previous qualitative analysis. Its principle is to simulate the occurrences of the primary events by a random number generator. In each trial, the primary event is simulated by generating a random number in the interval [0, 1], and if the number is no larger than the probability assigned, the event is reckoned to occur. Then the fault tree is evaluated for the top event probability and the contributions of the primary events by a large number of trials. In this analysis, the primary concern is the

TABLE 6.6 Probability of Occurrence for Damage Causes

Damage cause	Number of occurrence per flight hour (probability)
D1 surface protection	0.05
D2 mishandling	1.2e-4
D3 lightning strike	0.5e-4
D4 heat	0.5e-4
D5 material resistance	0.1 for dent/0.005 for hole
D6 natural object impact	0.8e-4
D7 wind erosion	1.2e-4
D8 manufacturing defects	0.02
D9 overloading	0.5e-4

probability importance of every basic event instead of the probability of the top event. Monte Carlo simulation was performed and the number of trials was set to 1e + 6; a table of the failure contribution towards the top event and the importance value of each basic event (damage cause) are obtained as discussed in the following section.

6.6 DISCUSSION

For quantitative analysis, current statistical damage data obtained from the survey is still not comprehensive. Some data needs to be either idealized or hypothesized based on engineering experience, such as the probability distributions of C1 Scratch, D1 Surface protection, D5 Material resistance, and D8 Manufacturing defects.

D5 Material Resistance is one of the inherent properties of composites, its resistance to low energy impact causing dent is weak whereas the resistance to large energy impact causing hole is relatively strong. Therefore, two discrete values are assigned to D5 for these two situations.

It should be noted that in Table 6.7 numerical values of failure contribution less than 1e-5 are neglected due to the program precision and thereby the corresponding D3, D7, and D9 with very low importance are set to 0. Overall, the importance ranking of damage causes is D5 > D8 > D2 > D6 > D4 = D1 > D3 = D7 = D9.

Previous qualitative analyses rank the importance of every basic event from three aspects. Structure importance analysis is based on the fault tree structure itself. Probability importance analysis reflects the unreliability of the basic event to the top event. Relative probability importance analysis was performed as a supplement measuring sensitivity. Rankings of the damage causes were obtained in Table 6.3 and Table 6.4. Compared with the results of the

TABLE 6.7 Importance of Damage Causes

Damage cause	Failure contribution	Importance
D1 surface protection	5.170e-5	9.66
D2 mishandling	3.057e-4	57.10
D3 lightning strike	<1.000e-5	0
D4 heat	5.170e-5	9.66
D5 material resistance	6.634e-5	123.81
D6 natural object impact	1.984e-4	37.06
D7 wind erosion	<1.000e-5	0
D8 manufacturing defects	4.482e-4	83.71
D9 overloading	<1.000e-5	0

numerical example, excluding the particular group (D1, D5, and D8) mentioned in the previous section, the importance ranking for the damage causes is D2 > D6 > D4 > D3 = D9 = D7, which is consistent with inequality Eq. (6.7), demonstrating the feasibility of the FTA on composite damages.

The benefits of the method are summarized as follows:

- A wide variety of composite damage modes and damage causes can be synthesized into a tree that is intuitive and systematic.
- Without sufficient information on damages, qualitative analysis can be performed to identify main contributors and then targeted actions and resources can be prioritized.
- With sufficient data available, quantitative analysis can be conducted. Either constant probabilities or time-related probabilities can be calculated to obtain the top event frequency, occurrence rates of damages, damage severity, and so forth, providing valuable information to maintenance and reliability departments.

6.7 POTENTIAL SOLUTIONS

According to both qualitative and quantitative fault tree analyses on CFRP composite damages, contributions of various damage causes have been prioritized. This method can be used as a proactive tool to prevent the occurrence of the top event from those main contributors. Several solutions addressing different damage causes are proposed in order to improve the reliability of composite structures.

6.7.1 Material Design

To improve the poor material resistance (D5) to impacts, great efforts should be paid to developing 3D composites, which can not only enhance through-thickness resistance, but also prevent from delamination propagation [51]. Typical examples are Z pinned composite and 3D fiber structures as shown in Fig. 6.3. However, there is still a long way to go before 3D structures are widely used by aircraft industries due to cost and efficiency. Economic manufacturing processes and new airworthiness regulations specific to 3D composites should be developed at the same time.

6.7.2 Fabrication Process

Different from traditional metallic components manufacturing, there are various forming processes for composites, such as autoclave forming, vacuum bag molding, pultrusion, filament winding, and resin transfer molding. After forming, machining is applied including cutting, trimming, drilling, and reaming [57]. Inherent flaws like voids, filament spacing, misalignment, imperfect interface bonding, residual stress, and so forth are introduced occasionally during the fabrication process.

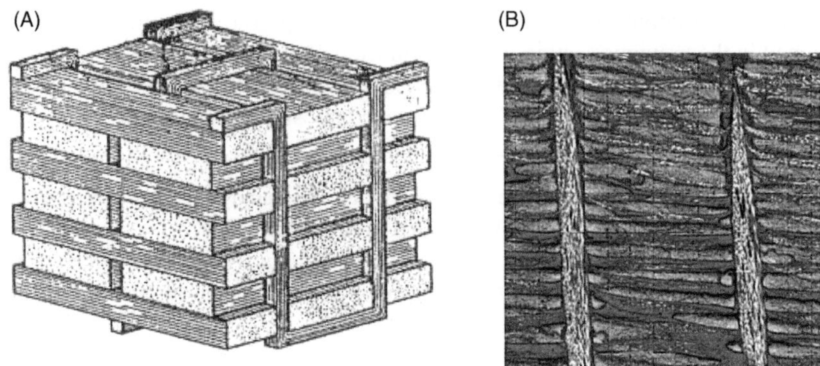

FIGURE 6.3 (A) 3D fiber structure and (B) Z-pins composite.

To reduce the manufacturing defects (D8), more accurate manufacturing process should be implemented. New techniques, such as a drilling method that can prevent laminate from edge fuzz and an exact temperature control in autoclave forming can be developed. Meanwhile, a more strict quality certification procedure should be applied to enhance manufacturing quality control.

6.7.3 Personnel Training

Since accidental damage from natural sources is hard to predict, great attention should be paid to the improvement of technical skills of the operating personnel to reduce the human mishandling (D2). From the previous analysis, human mishandling (D2) makes a significant contribution in damage threats, such as ground vehicle collisions, tools dropping, and so forth. The qualification required to maintain composites is much higher than that to maintain traditional materials. Targeted training procedures should be further studied and implemented. Maintenance workload should be reduced and the working environment should be improved to avoid unnecessary mistakes.

6.7.4 Surface Protection

Adequate surface protection (D1) is one of the key factors in scheduled maintenance. Efforts should be put to investigating main damage causes occurring in different locations of aircraft so that targeted protective coating can be applied to effectively reduce specific damages due to lightning, erosion or moisture. The development of new coating techniques with multiple protective functions is encouraged.

6.7.5 Damage Evaluation and Life Prediction

Except for inherent reasons (D1, D5, and D8), impact damage is the most significant cause of the composite structural failure. Compared to the understanding of metal crack propagation due to fatigue, the deterioration for composites after impact is yet to be determined. Investigations into the mechanisms of composite damage accumulation should be continued to characterize the relationship between the size of the impacted area and the residual strength in order to make more accurate life prediction. Then, optimized inspection intervals can be determined to monitor the composite structural health, satisfying both safety and economic requirements.

6.8 CONCLUSIONS

This chapter proposed a new FTA to synthesize a diversity of damage modes and damage causes of the composite structure in a systematic manner on a macroscopic level. A typical composite panel made of CFRP was selected and three importance analyses including structural importance analysis, probability importance analysis, as well as relative probability importance analysis were conducted to rank various damage causes. The applicability of the FTA on composites was validated through a numerical example based on statistical data and engineering experience from survey. Potential solutions aiming at improving the reliability of composite structures were proposed accordingly. Engineers from airlines can apply this method to discover the main damage modes and damage causes for different composite structures through operational monitoring so that pertinent preventative actions can be performed. Manufacturers can combine this approach with other methodologies, such as fuzzy comprehensive evaluation, back-propagation network, and so forth to develop composites' rating system for more efficient maintenance schedules.

of mechanical properties due to fatigue, the determination for compression after impact is yet to be determined. Investigations into the mechanisms of [illegible] the damage accumulation should be continued to characterize the relationship [illegible] [illegible] compared to [illegible] structural health [illegible] [illegible].

[illegible]

[illegible]

[illegible]

Chapter 7

Inspection Interval Optimization for Aircraft Composite Structures Considering Dent Damage

Chapter Outline

7.1 INTRODUCTION

Currently, aircraft structures are designed based on deterministic approaches, which assume a worst-case scenario and give a factor of safety for the loads and a knockdown factor for the strength. One of the fundamental disadvantages is that the deterministic approach is developed under conventional configurations, mostly for metallic materials and familiar structural designs. The determination of those factors comes from engineering experiences in metallic aircraft in the past half century. However, new aircraft design philosophies are developing dramatically away from conventional environments. Historical uncertainty factors may not be able to provide sufficient safety and reliability. Designing to all

http://dx.doi.org/10.1016/B978-0-12-812668-4.00007-1

worst-case scenarios may result in an unacceptable increase in weight. In addition, new materials also pose challenges for current structural design. The use of composite materials in aircraft structures is growing because of many advanced properties, such as high strength/stiffness ratio, low sensitivity to environment, design flexibility, and so forth. But these materials have more intrinsic variables than metals due to heterogeneity and they are subjected to more sources of variation in the manufacturing process [58]. The extra uncertainties lead to relatively large knockdown factors, which also result in substantial weight increase without a quantifiable increase in structural reliability.

The other disadvantage of the traditional design approach is a lack of quantifiable measurement for safety and reliability. Therefore, it is difficult to select an optimized design approach on aircraft safety as designs are becoming more critical and competitive. Both military and commercial guidance emphasize the importance of reliability as an essential feature. With numerical reliability values available during operation, a consistent level of safety and efficiency can be ensured throughout the aircraft life cycle.

New design philosophies and new materials have introduced more uncertainties and there is a need for maintaining life cycle efficiency with measureable indexes. Therefore, nondeterministic design methods arise as an alternative but advanced solution to address uncertainties, in other words, to reduce potential risks.

Risk is defined as the possibility of encountering harm or loss, and it exists in all activities [58]. People decide to take part in a particular activity and accept the risk level, either consciously or subconsciously. For aircraft structures, risks come from various factors, from design to operation since most of these engineering parameters have a random nature, such as material properties, working environment, loading, damage, and so forth. Different from the traditional design analysis that treats them as deterministic values and therefore leaves an unknown reliability, probabilistic structural analysis provides a means to quantify the inherent risk of the design.

In terms of aircraft maintenance, pertinent activities are determined mainly based on two factors: (1) how aircraft is designed and manufactured and (2) the operational conditions. For an aircraft structure, the maintenance planner needs to know the load-bearing capacity of the structure and the structural degradation mechanism to determine various possible damages during operation. Thus, risks in maintenance can also be quantified by probabilistic structural analysis methods, since in addition to design variables, damage characteristics, and operational environments are probabilistic in nature. The uncertainties exist almost everywhere during the life cycle of the aircraft structure. The probabilistic analysis is intended to address all cases in a practical approach instead of considering them in the worst-case scenarios. With sufficient design and service data made available, maintenance plans can be optimized by reducing the lifelong cost while maintaining an acceptable risk level that is intuitively measurable.

Currently, the MSG-3 document is the most widely accepted method for creating maintenance tasks and intervals. However, even though an intelligent rating system was proposed to determine more objective inspection intervals, it still largely depends on engineering experience without considering a variety of factors from design, manufacturing, and operation. Besides, its rating ability for new generation materials is relatively limited, which makes it insufficient for new aircraft with a significant amount of advanced composite structures.

This chapter introduces a probabilistic approach to address aircraft composite structures, which is beyond the scope of the experience-based MSG-3 method. Instead, it is driven by data combined with specific physical properties, which is more reliable in determining optimized inspection intervals for aircraft composite structures.

7.2 DAMAGE TOLERANCE PHILOSOPHY OF COMPOSITE STRUCTURES

7.2.1 Properties of Aircraft Composite Structures

Composite properties have been discussed in the previous chapters. Because of their anisotropy and complicated manufacturing processes that are difficult to control precisely, composite structures display more complex damage modes compared with cracks in metallic structures. Fortunately, in terms of three damage sources, historic service experience shows that composite primary structures that comply with FAR-25 and JAR-25 requirements, such as Boeing 737 horizontal stabilizer, Boeing 777 and Airbus series empennage, and so forth exhibit excellent resistance to environmental deterioration and fatigue damage [59–61]. Therefore, for primary structures, such as thick skins, accidental damage becomes the primary concern in damage tolerance design and maintenance planning.

7.2.2 Maintenance Model of Composite Structures

Generally, metallic materials are homogeneous and have high ductility. A clear fatigue characteristic can be obtained and thereby crack propagation in metallic structures can be predicted precisely. In comparison, composite materials show brittle behavior during failure because of their anisotropy and large scattered properties. From previous fault tree analysis, composites are much more sensitive to impact damage, but there is no general damage propagation mechanism up till now. Therefore, composite structures are designed to preclude detrimental damage growth during normal operation. To be more specific, composite structures follow a "no-growth" approach to damage-tolerant design. All composite SSIs are certified to meet the no-growth requirements. Damage initiation and propagation due to fatigue is not applicable for composites [62,63]. Damage in composites is assumed to stay constant instead of deteriorating with time. A typical figure is shown in Fig. 7.1 [64].

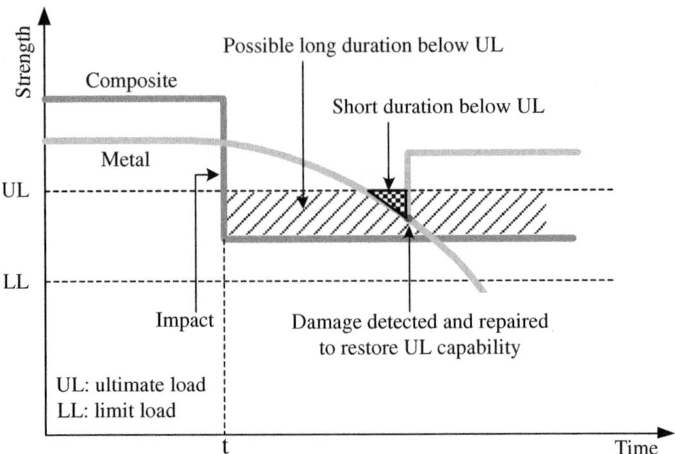

FIGURE 7.1 Damage tolerance design for metals and composites.

Two kinds of load conditions are defined by FAA [65]:

1. Limit loads are the maximum loads expected in service and there should be no permanent deformation of the structure at limit load.
2. Ultimate loads are defined as the limit loads times a safety factor. In FAR Part 25 the safety factor is specified as 1.5.

For different aircraft usage, the safety factor can be adjusted. The structure must withstand the ultimate load for at least 3 seconds without failure. Usually, there are two situations in which the safety factor is more than 1.5. One is for some very critical load-bearing components in order to guarantee a high safety level. The other is for structures made of composite materials because of their brittle behavior.

As depicted by the yellow curve in Fig. 7.1, assume a crack initiates at time t in a metallic structure. The residual strength goes through a degradation process as the crack grows with time; once the damage is beyond the allowable size, the structure must be repaired. Otherwise, the strength will further degrade. If the residual strength decreases below the limit load level, severe safety problems might be induced. In comparison, assume an impact occurs at time t on a composite structure, as depicted by the blue curve. The residual strength immediately reduces to a lower level but still stays above the limit load level. According to the "no-growth" design philosophy, the damage size will not grow further regardless of the cyclic loading. In this situation, it seems that the damage tolerance assumption can always be met without the need to repair. However, there is another hidden but inherent parameter, which is difficult to measure by deterministic methods: the probability of failure of the damaged composite structure will increase, as depicted by the shaded area in Fig. 7.1. If the damage in the

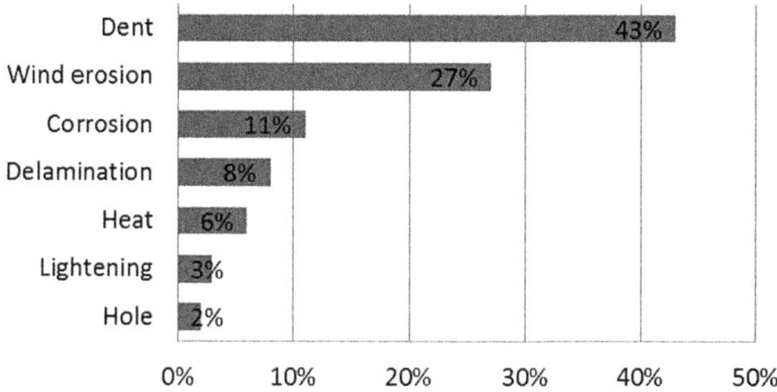

FIGURE 7.2 **Damage category.**

metallic structure is detected and repaired in a timely way, whereas the damage in the composite structure remains undetected or unrepaired for a long time, the composite structure will not be as safe as the metallic one. This is why an inspection schedule and a repair policy are necessary under the damage tolerance philosophy of composite airframes.

7.3 DAMAGE CHARACTERIZATION

The prerequisite of the probabilistic approach is to obtain necessary information in terms of composites design properties and actual usage during operation. A survey was conducted at an airline maintenance department to collect damage records on aircraft structures made of composite materials. Then statistics were performed to refine valuable information as inputs.

7.3.1 Data Statistics and Category

Data on wing structural damage (ATA Chapter 57) for two types of aircraft fleet (Boeing 737-800 and Boeing 757-200) over the past 10 years were obtained. A typical damage category of composite structures is shown in Fig. 7.2. It should be pointed out that although paint peel-off is often mentioned throughout the maintenance records, it is simply a superficial defect that will not cause the loss of inherent integrity, and thereby not listed.

The bar chart illustrates that impact damage consisting of dent, delamination, and hole are responsible for more than 50% of all damages, and dent is the primary damage type. Different from the internal damages like delamination and hole, dent is commonly caused by various discrete low-energy impact sources, such as hail, runway debris, tools dropping, and so forth. This chapter focuses on the low-energy impact damage resulting in dent, and the effect of dent sizes.

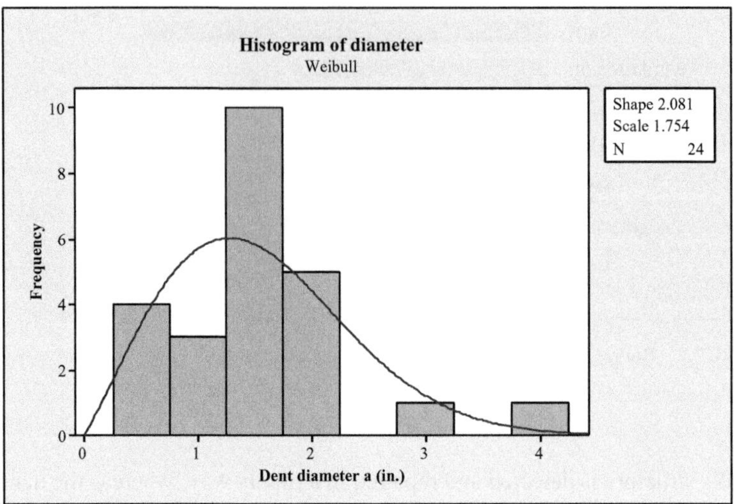

FIGURE 7.3 **Damage diameter distribution.**

7.3.2 Damage Size Distribution

The appearances of the dent can be superficial with a large surface area or deep in the material with a small surface area. Statistical analysis was performed on damage area diameter a and depth y. A goodness-of-fit test was performed for several probability models in order to determine the best fitness to describe the damage diameter and depth. Results show that the classic Weibull distribution has a good fitness with a correlation coefficient of 0.969 and 0.981, respectively. The probability distributions for diameter and depth are shown in Fig. 7.3 and Fig. 7.4.

The Weibull expression of the diameter distribution is given by:

$$f(a;\alpha,\beta)=\frac{\beta}{\alpha^{\beta}}a^{\beta-1}e^{-\left(\frac{a}{\alpha}\right)^{\beta}},\quad a\geq 0,\quad \alpha>0,\quad \beta>0$$
$$\alpha=1.754,\quad \beta=2.081$$

The Weibull expression of the depth distribution is given by:

$$f(y;\alpha,\beta)=\frac{\beta}{\alpha^{\beta}}y^{\beta-1}e^{-\left(\frac{y}{\alpha}\right)^{\beta}},\quad y\geq 0,\quad \alpha>0,\quad \beta>0$$
$$\alpha=0.05636,\quad \beta=2.456$$

Inspired by Ren [66], who proposed a corrosion-spot index (CSI) to describe the relationship between the depth and the area of a corrosion spot, an

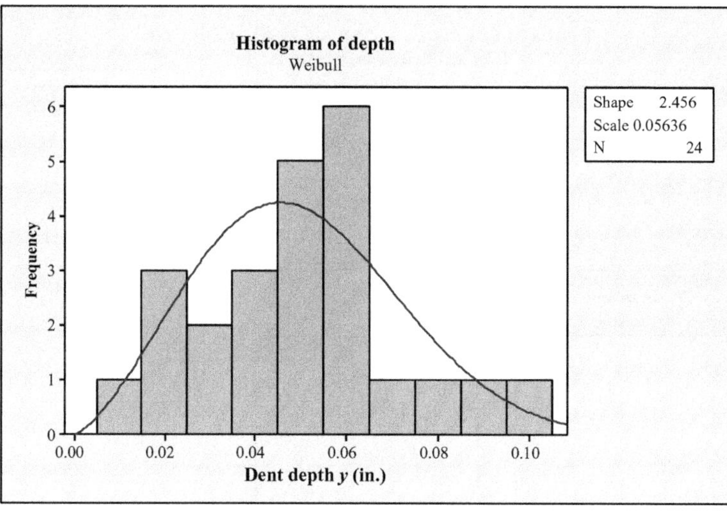

FIGURE 7.4 **Damage depth distribution.**

analogous concept called dent spot index (DSI) is proposed, which denotes the diameter/depth ratio and is expressed as:

$$\eta = \frac{a}{y}$$

Four most likely probability distribution functions (PDFs) were tested and their goodness-of-fit is calculated in Fig. 7.5.

It is apparent from the test that log-logistic distribution function is the best PDF with the highest coefficient 0.981. Its scale and location parameters are shown in Fig. 7.6.

The log-logistic expression of this DSI distribution is given by:

$$f(\eta;\alpha,\beta) = \frac{(\beta/\alpha)(\eta/\alpha)^{\beta-1}}{\left[1+(\eta/\alpha)^{\beta}\right]^2}, \quad \eta > 0, \quad \alpha > 0, \quad \beta > 0$$
$$\alpha = 0.3118, \quad \beta = 3.427$$

The structural repair manual (SRM) of the Boeing 737-800 has similar definitions [67]. For example, one of the allowable damage limits for a damage area is specified as: w/y must be 30 or more at each point along the length of the dent and y = a maximum of 0.125 in., where w = minimum width of the dent and y = depth of the dent where the width is measured.

DSI may provide more practical values in damage detection and in the subsequent repair decision making considering both damage diameter and depth. Unfortunately, there is little information on the effectiveness of probability of

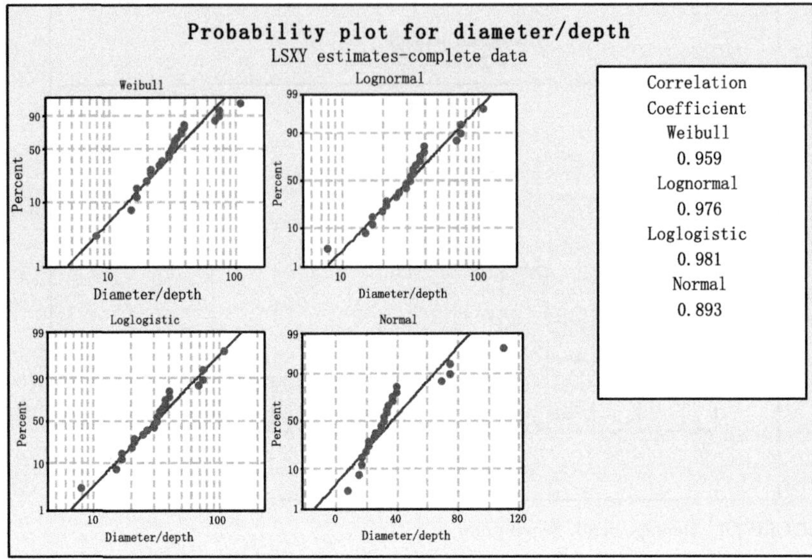

FIGURE 7.5 **Goodness-of-fit test for DSI.**

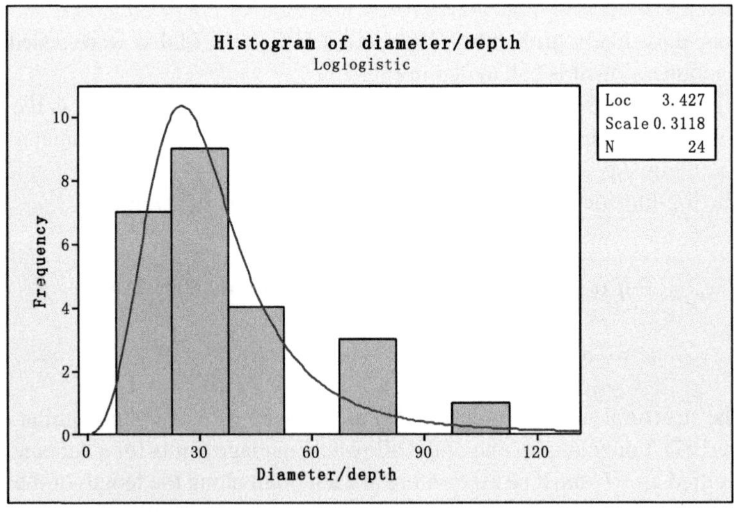

FIGURE 7.6 **Diameter/depth distribution.**

detection (POD) versus DSI. In this case, the PODs for damage area diameter and depth are studied and the larger one is selected as the final POD.

7.3.3 Probability of Detection (POD)

It is a time-consuming effort to perform experiments to examine engineers' possibility to detect dent with influencing factors incorporating colors of the composite panel, cleanness, inspection duration, brightness, personal eyesight, professional skills, the inspection angles, and so forth. [68]. General visual inspection (GVI) and detailed inspection (DET) are considered as two general inspection types in structural maintenance. Special detailed inspection (SDI) is applied to make more detailed examination. Previous studies focused on POD for detecting cracks in metal structures and composites show the similar results [69], which are demonstrated by some recent experimental results on certain composite panels [68]. The POD values generally follow two distributions and their cumulative probability functions: Two equations can be used as approximations without data scale.

Cumulative Weibull distribution function:

$$F(x)=1-e^{-\left(\frac{x}{\beta}\right)^{\alpha}} \tag{7.1}$$

Cumulative log-normal distribution function:

$$F(x)=\frac{e^{\alpha+\beta\ln(x)}}{1+e^{\alpha+\beta\ln(x)}} \tag{7.2}$$

In terms of damage area depth, experimental data was obtained for a green panel at 45-degree inspection angle by DET [68]. Meanwhile, the POD data for GVI was obtained according to engineering experience in the surveyed maintenance department. The POD curves and the related parameters are shown in Fig. 7.7 and the following two equations.

$$\text{POD\_GVI}(y)=1-e^{-\left(\frac{y}{\beta}\right)^{\alpha}},\quad \alpha=3.318,\quad \beta=0.03732 \tag{7.3}$$

$$\text{POD\_DET}(y)=1-e^{-\left(\frac{y}{\beta}\right)^{\alpha}},\quad \alpha=2.896,\quad \beta=0.01921 \tag{7.4}$$

In terms of damage area diameter, data is obtained from [70] and it shows that log-normal model fits the data with a higher regression value. The POD curves and the related parameters are shown in Fig. 7.8 and the following two equations.

$$\text{POD\_GVI}(a)=\frac{e^{\alpha+\beta\ln(a)}}{1+e^{\alpha+\beta\ln(a)}},\quad \alpha=-5.619,\quad \beta=7.352 \tag{7.5}$$

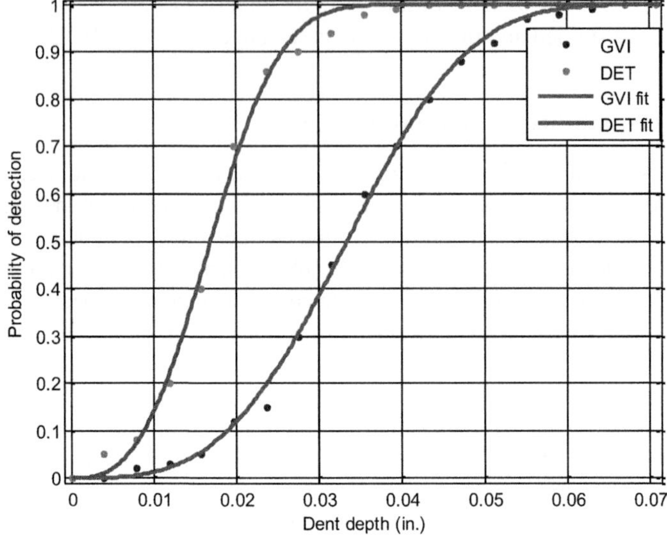

FIGURE 7.7 POD versus damage depth (in.).

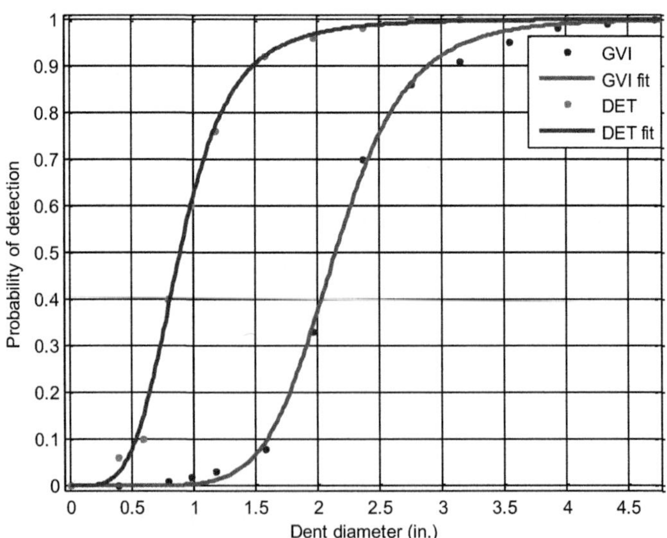

FIGURE 7.8 POD versus damage diameter (in.).

$$\text{POD\_GVI}(a) = \frac{e^{\alpha+\beta \ln(a)}}{1+e^{\alpha+\beta \ln(a)}}, \quad \alpha = 0.4877, \quad \beta = 4.294 \tag{7.6}$$

7.4 PROBABILISTIC METHOD

7.4.1 Reliability Formulation

One of the objectives of applying probabilistic damage tolerance philosophy is to be able to quantify the reliability of the structure. Herein, the assessment of the probability of failure (POF) is illustrated as follows:

According to the requirement of damage tolerance philosophy, damage accumulated in service of a structure should be detected and repaired before the residual strength degrades beyond some predetermined threshold. In this study, the entire service life cycle of a composite structure is considered as a series of discrete activities consisted of damage, inspection, and repair (a life cycle denotes the time period from the aircraft entering into service to its retirement). These activities exerted from outside can be reflected by the variation of the inherent residual strength of the structure. The POF per life cycle is evaluated by the following formulation:

$$\text{POF} = 1 - \prod_{i=1}^{N} \left[1 - P_f\left(S_i, t_i\right)\right] \tag{7.7}$$

where t_i is the ith time interval between (i–1)th and ith activity (0 means the initial service time), S_i is the ith residual strength between (i–1)th and ith activity, N is the number of damages occurred in one life cycle, and $P_f(\cdot)$ is the probability of failure for each interval with constant residual strength.

Failure occurs when the applied load exceeds the residual strength. Each time interval throughout the life cycle with constant residual strength is assumed in series connection. The cumulative distribution function (CDF) of the maximum load per t_i is expressed as:

$$F_l\left(S_i, t_i\right) = e^{-H(S_i)t_i} \tag{7.8}$$

where $H(x)$ is the frequency of the event exceeding the level x.

A simple numerical example is provided later.

Assume that a composite panel has a random residual strength history as shown in Fig. 7.9.

The initial strength is 1.5. An impact damage occurs at the instant $t0$. The residual strength decreases to 1.2 and stays constant until the time $t1$ when the damage is detected and the strength is recovered to its original level immediately. After $t1$, there is no damage to this structure till the end of life. Failure occurs when the random external load exceeds the residual strength. The load exceedance curve is assumed in Fig. 7.10 and is expressed as:

$$H(x) = 10^{10} e^{-23.03x}$$

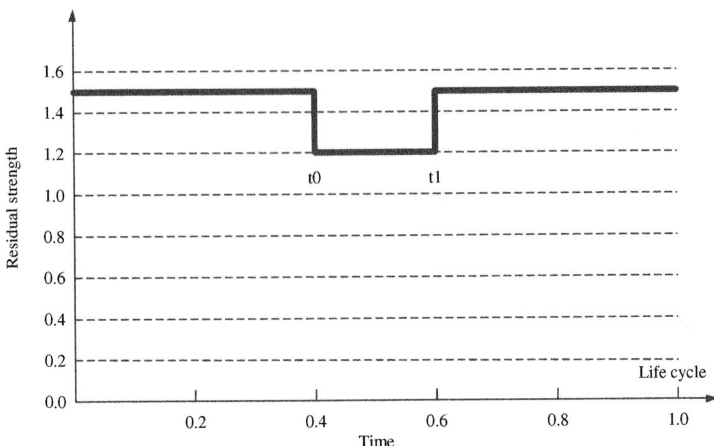

FIGURE 7.9 Random residual strength history in a life cycle.

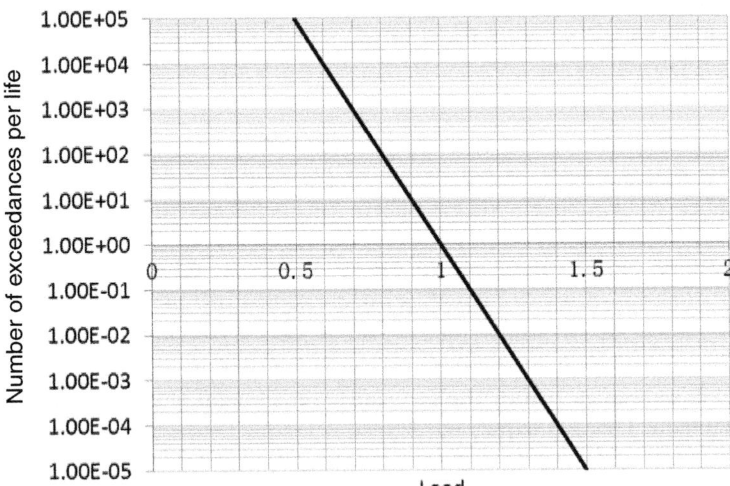

FIGURE 7.10 Load exceedance curve.

There are three time intervals t_i, with a constant strength S_i as shown in Fig. 7.9. Take the first time interval [0, 0.4], for example:

$$\begin{aligned} P_f(S_1, t_1) &= 1 - F_1(S_1, t_1) \\ &= 1 - e^{-H(S_1)t_1} \\ &= 1 - e^{-H(1.5)\times 0.4} \\ &= 3.975 \times 10^{-6} \end{aligned}$$

Likewise, $P_f(S_2,t_2) = 1.988\times10^{-3}$ and $P_f(S_3,t_3) = 3.975\times10^{-6}$. Then the final POF value can be obtained by Eq. (7.7).

The calculation of POF depends on a number of random variables, such as the number of damage occurrence, damage occurrence time, the damage extent, and other factors including loading conditions, inspection schedule, and residual strength, to name a few. Hereafter, a probabilistic simulation procedure based on the Monte Carlo method is proposed, incorporating all the influencing parameters necessary to evaluate the POF.

7.4.2 Monte Carlo Simulation

Monte Carlo simulation is a computerized mathematical technique that models phenomena with significant uncertainty. This method provides an effective means to account for risk in quantitative analysis and decision making. It performs risk analysis through building models of possible substitution of a range of value sampling from probability distributions, avoiding costly and time-consuming experimental repeats [71].

Monte Carlo can be used to model both discrete and continuous systems. Take discrete system as an illustration. The system state is driven by random events at limited/countable time. The state only changes at discrete random time moments and it is assumed that the change is completed instantly.

The entire life cycle of a composite structure suffering from low energy impact can be considered as a series of discrete activities incorporating damage, inspection, and repair. Due to the no-growth damage tolerance philosophy, the residual strength stays constant until being repaired. Therefore, the residual strength variation in a life cycle can be considered as a discrete system. Monte Carlo simulation has proven to be a robust methodology for such complicated problems with discrete random variables [72]. It generally tends to follow a particular pattern:

- Define a domain of possible inputs
- Design a logic block diagram
- Generate inputs randomly from probability distributions over the domain
- Perform a deterministic computation on the inputs

The flowchart of the logic procedure is described in Fig. 7.11.

7.5 CASE STUDY

7.5.1 Average Damages Per Life Cycle (*Nd*)

Information on dent was extracted from the maintenance records of the Boeing 757-200 fleet wing structures (ATA, Ch. 57) from 2002 to 2012. In total, 46 occurrences of dent damage were recorded in six aircrafts, among which 12 occurred on 4 GFRP panels located near the wing leading edge. The design life (life cycle) of the Boeing 757-200 is 150,000 flight hours and these composite panels are assumed to have the same design life. Thus, the number of dent events per life cycle on average (*Nd*) is 2.5. This is a relatively rare event, and *Nd* is described by a Poisson distribution.

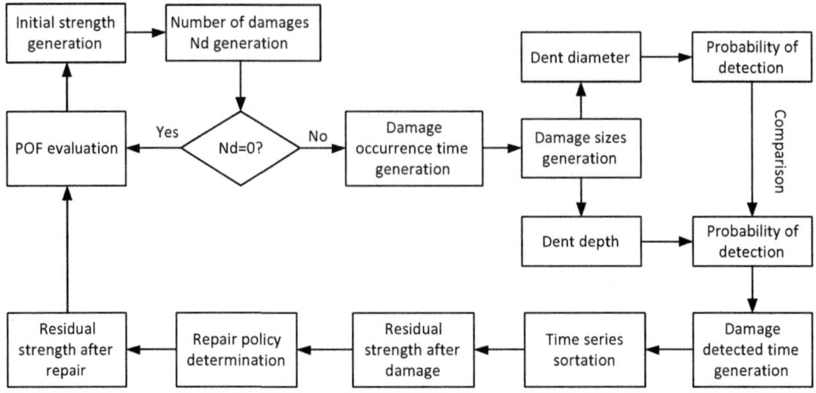

FIGURE 7.11 Simulation flowchart for dent damage.

7.5.2 Load Cases

It is difficult to obtain the specific load exceedance data for certain structures. But for civil aircraft, the occurrence of gust load is mainly considered [73]:

$$P_{\text{rat}} = P(\varepsilon > \varepsilon_{\text{LL}}) = 2\times10^{-5}\,/\text{FH},$$
$$P_{\text{rat}} = P(\varepsilon > \varepsilon_{\text{UL}}) = 1\times10^{-9}\,/\text{FH}$$

Since the probability value under different exceeding conditions changes significantly by the power of 10, a Log-linear model is used to describe the load occurrence probability. The load exceedance curve is shown in Fig. 7.12 and is expressed as:

$$\lg(P_{\text{rat}}) = -8.602\times\left(\frac{\varepsilon}{\varepsilon_{\text{LL}}}\right) + 8.903$$

7.5.3 Damage Size and Occurrence Time

The damage occurrence times are a series of highly random variables throughout the entire service life. A uniform generator was used to scatter the operational damage in one life cycle. Generally, we use the damage area diameter and depth to describe a dent. In reality, some damage caused by a sharp object may be deep with a small area whereas some damage caused by a blunt object may be shallow but has a considerable damage area and even delamination. Accordingly it is difficult to describe their relationship. Therefore, two generators were used to generate the dent diameter and depth, obeying the Weibull PDF with separate parameters.

Theoretically, the damage size is a function of many variables, such as the quality of the manufacturing process, the thickness of the laminate, the size

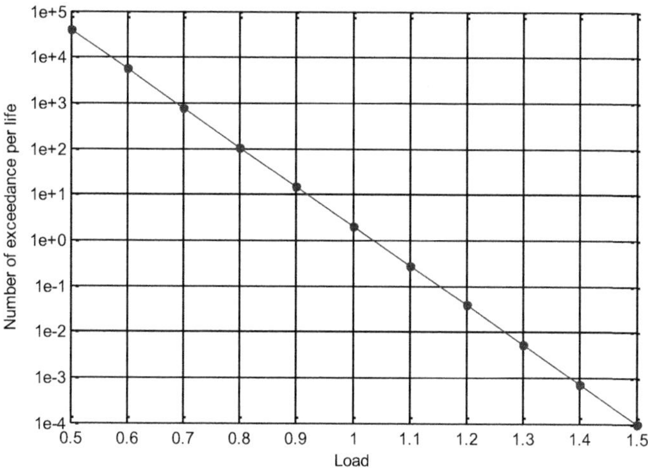

FIGURE 7.12 Load exceedance curve of gust load.

of the impact object, impact energy, load case, boundary condition, composite design, and so forth. Many studies have considered different decision variables either in theory or by experiments. However, during real operation, the maintenance engineers do not consider, for instance, the boundary condition or the impact energy. They work only with what they see by eye or with instruments: the damage size. Therefore, this study focuses on damage size data including damage diameter and depth and also the DSI.

7.5.4 Inspection Efficiency

The inspection efficiency is described by the probability of detection (POD), which was explained in the previous section. The inspection interval is preset as T and the random time to detect the damage is expressed as:

$$t = T \times \xi,$$

where ξ is the number of inspection times to detect the damage, which can be generated by a geometric distribution [74]. Assume the jth damage occurs at t_j, then is detected at the ξth inspection after the damage, which can be expressed as:

$$t_d(j) = \left(\left[\frac{t_j}{T} \right] + \xi \right) \times T,$$

where $[\cdot]$ is the floor operator.

7.5.5 Residual Strength Reduction and Recovery

In order to calculate the probability of failure, the damage size must be converted to the reduction of residual strength. For the most severe load case, the

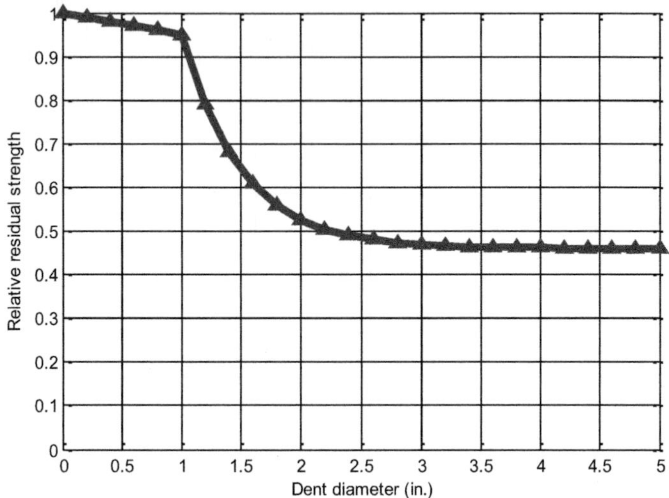

FIGURE 7.13 **Relative residual strength reduction.**

compression capacity is mainly considered and the damage diameter is used as the decision variable. The relative strength of the damaged GFRP panel is described by the following function:

$$RS(a) = \begin{cases} 1 - ka, & 0 < a < 1, \quad k = 0.05 \\ A + (C - A)e^{\left(\frac{a}{G}\right)}, & a \geq 1, \quad A = 0.46, \quad C = 4.08, \quad G = 0.5 \end{cases} \tag{7.9}$$

where A is the residual strength asymptote; C is the intercept; G and k are the slopes for the two curves. The relative strength reduction curve is shown in Fig. 7.13.

The recovery of the residual strength depends on different repair policies. Once damage is detected, engineers should refer to the detailed criteria in the SRM to decide whether the damage should be left as it is or be repaired (replaced). For the sake of simplicity, we assume the following rules: dents with diameters less than 1 in. can be allowed whereas larger dents must be repaired to recover to its r% strength, where r is the recovery efficiency coefficient described by a uniform distribution within the range of 0.85–0.95.

7.5.6 Other Assumptions and Definitions to Facilitate the Simulation

The initial strength is described by a Gaussian distribution with the coefficient of variation 5%, referring to the fact that many strength analyses apply a Gaussian PDF [75]. The initial average value of the residual strength is

$$\mathrm{RS} = f \times f_1 = 1.5 \times 1.4 = 2.1,$$

where $f = 1.5$ is the factor of safety and $f_1 = 1.4$ is the additional margin of safety.

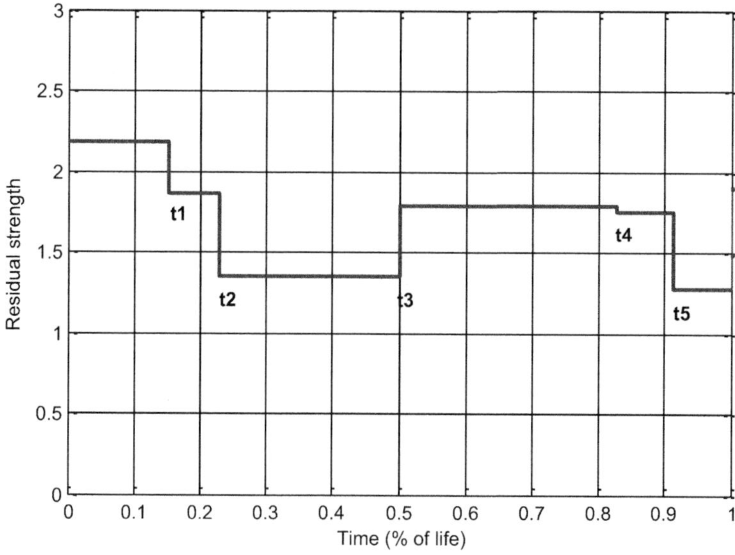

FIGURE 7.14 Residual strength per life cycle.

Environmental factors, such as temperature, moisture and ultraviolet light may cause deterioration of composite materials. Also, adjacent damage sites in a composite panel may result in additional strength reduction. For the sake of simplicity, these factors are not considered. Also, the repair duration is ignored, compared to the long life cycle operation. Three situations are assumed:

1. Damage is not detected. No repair activity is required;
2. Damage is detected and within the threshold. No repair activity is initiated;
3. Damage is detected and beyond the threshold. Repair is required.

7.6 SIMULATION RESULTS AND DISCUSSION

Set the life cycle to 1 unit and the inspection interval $T = 0.1$, the inspection method is GVI. A typical example of the residual strength in a random life cycle is plotted in Fig. 7.14.

Apparently, there are four damage occurrences on the composite panel in one life cycle. At the instant $t1$, the panel suffers the first impact but the damage is not detected in the following GVI inspections. The second damage occurs at $t2$ and it is detected and repaired at $t3$. The third damage occurs at $t4$, which is either undetected or detected but reserved. The last damage occurs at $t5$, but before any inspection begins, the structure reaches its end.

Various life-cycle strength results can occur due to the randomness of the Monte Carlo simulation. By taking 1000 samples, we obtain the relationship between the average probability of failure (POF) and the inspection interval

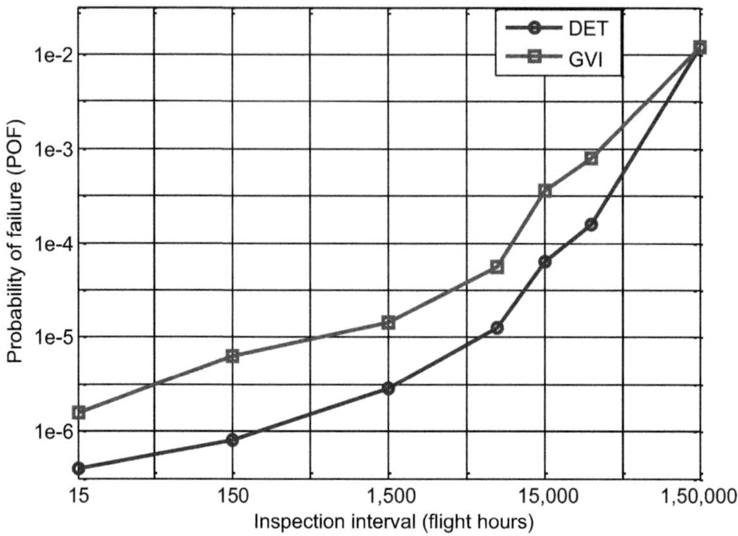

FIGURE 7.15 POF versus inspection interval.

T by GVI and DET, respectively; see Fig. 7.15. (Both axes were processed by logarithm, same for Fig. 7.16).

It is shown that fewer inspections result in a higher probability of failure. By assigning each inspection interval a risk level, here described by a POF value, airline engineers can determine the inspection intervals by assessing the required reliability of the composite structure due to different structural configurations in different service situations. For instance, if the required POF is no higher than 10^{-4}, the maximum inspection intervals by GVI and DET should be 9,600 flight hours and 21,000 flight hours, respectively.

Economy is the second important factor next to safety for civil aircraft. The total in-service cost for maintenance of the composite structure is classified into three parts:

1. Routine inspection cost for the structure;
2. Repair cost of the damage detected in the inspection;
3. "Penalty cost" due to the structural high failure risk.

Assumptions are as follows: set the cost for each inspection by GVI to 1 unit and inspection by DET to 5 units; if the damage is detected and repaired, the repair cost is in proportion to the reduction of the residual strength; "penalty cost" is induced when the POF is above 10^{-4} level, in which case structures may be severely damaged, resulting in additional cost for replacement, spare parts, more labor hours, and so forth. The optimization of the inspection interval against the maintenance cost considering both safety and economy is shown in Fig. 7.16.

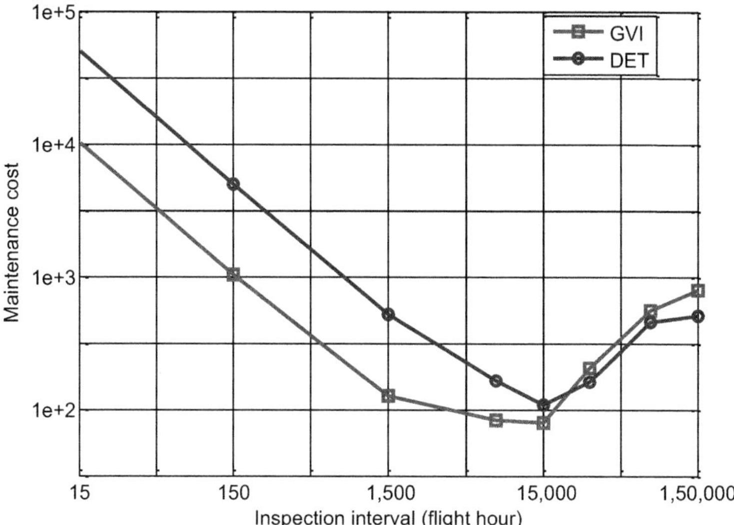

FIGURE 7.16 Inspection interval optimization.

By setting a POF threshold, the maintenance cost will not always decrease as the inspection interval increases. Because if the damage remains for a long time, the risk of failure will be high and therefore, the extra maintenance cost is very likely to be induced. It is shown in Fig. 7.16 that the minimum costs for both GVI and DET occur at approximately 15,000 flight hours. According to the MRBR, the inspection interval for the composite panel is "4C" (16,000 flight hours) by DET, which is the same level as the simulation result. The advantage of the methodology is that the airline can adjust their inspection intervals dynamically for different composite structures in different operational environments by setting an acceptable risk level to seek the most economical inspection schedule.

7.7 CONCLUSIONS

This chapter combined a data-driven technique with a physical model of a composite structure, and applied a probabilistic methodology beyond the limit of MSG-3, which is largely based on engineering experience. A probabilistic simulation procedure was established to describe the structural strength variation in a life cycle in order to optimize the inspection intervals in two criteria: maintaining a high structural reliability as well as minimizing the maintenance cost. A composite panel made of GFRP from in-service aircraft was selected to demonstrate the effectiveness of the methodology.

Aimed at in-service dent damage, the residual strength of the composite structure susceptible to low energy impact over a life cycle is simulated based on a "no-growth" design philosophy. By assigning each inspection interval a structural risk level and a cost factor, engineers from airlines and manufacturers can adjust the inspection intervals according to their specific requirements, satisfying both safety and economic objectives. Further, this method can be extended to include more factors, such as the impact of temperature, moisture, ultra-violet, and so forth. provided that sufficient data pertinent to the structural degradation mechanism is obtained. Last but not least, the probabilistic method developed in this chapter is quite flexible to be used in tackling many other problems, such as the repair issue, which is illustrated in the next chapter.

Chapter 8

Repair Tolerance for Composite Structures Using Probabilistic Methodologies

Chapter Outline

8.1 INTRODUCTION

In scheduled maintenance of aircraft structures, there are two key techniques: the determination of the inspection interval and the selection of the maintenance task [8]. Except for some general lubrication and servicing tasks, the major concern for damaged structures is whether to repair or replace. Specifically, during each inspection, damage detected should be evaluated by certain criteria to determine a suitable maintenance activity. This chapter deals with the repair issue in scheduled maintenance of composite structures.

The primary objective of structural repair is to restore the residual strength/stiffness of the damaged structure to its service condition in a limited time span and at a low cost [76]. Depending on the damage mode, damage location, and damage severity, structural repair can have different types. Therefore, damage assessment is an important step in selecting a specific repair activity. For aircraft composite structures, which are susceptible to impact damage but have good resistance to environmental deterioration and fatigue propagation, impact events often cause combinations of damage. For example, high-energy impact may

Reliability Based Aircraft Maintenance Optimization and Applications
http://dx.doi.org/10.1016/B978-0-12-812668-4.00008-3

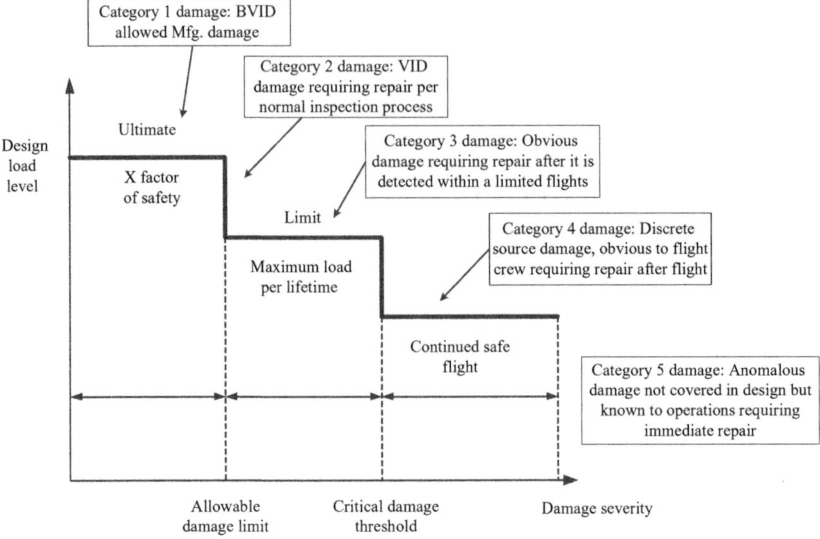

FIGURE 8.1 Design load levels versus categories of damage severity.

result in significant fiber breakage, matrix cracking, delamination, and broken fasteners; low-energy impact may include a combination of broken fibers, matrix cracks, and multiple delaminations. In some cases, damage may appear to be small on the surface but severe inside. Based on the damage tolerance design of composite structures for transport category aircraft, the airworthiness requires that catastrophic failure due to fatigue, environmental effects, manufacturing defects, or accidental damage should be avoided throughout the structural operational life cycle. Damages in composites are divided into the following five categories as depicted in Fig. 8.1 [77].

It can be found that except for Category 1 barely visible impact damage (BVID), which can be kept as it is, the other four damage categories require certain repair activities according to increasing damage severity. Category 2 and 3 belong to scheduled maintenance, since it is designed to relate the tasks to the consequences of structural damage remaining undetected. The remaining two categories requiring immediate repair are within the scope of unscheduled maintenance, which have much lower probability of occurrence compared to the other categories.

Once the damage is identified, maintenance personnel resort to source documentations to check with criteria. If the damage size is within the allowable limit, only simple maintenance work is needed, such as surface protection replacement, damage seal, and so forth. Otherwise, the damaged structure needs to go through a complex repair process. Sometimes the structure is even discarded and replaced with a new component. A category of various composite repair techniques is shown in Fig. 8.2 [78].

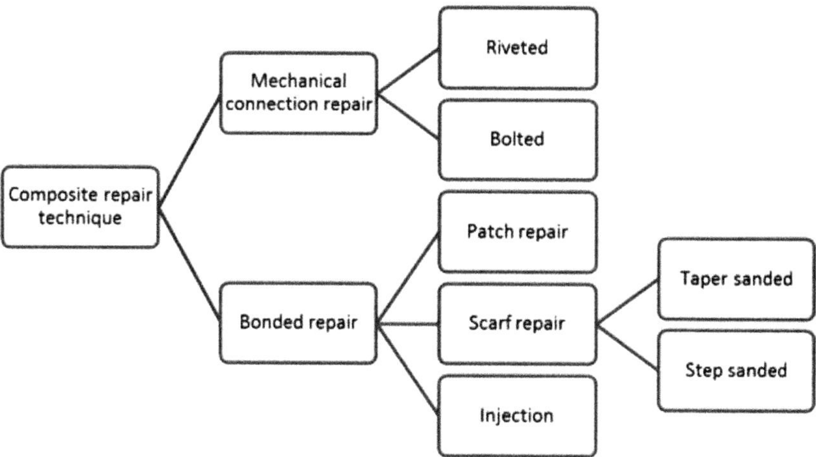

FIGURE 8.2 Composite repair techniques.

There are two major repair techniques, depending on the specific composite part and damage type: bolted repair and bonded repair. The concept of bolted repair is borrowed from conventional metal repair but with unique design and process details. Compared with bonded repair, it is simple and quick, for example, bolting a patch over the damage area, which can eliminate many potential problems induced in bonded repair. But bolted repair changes the original shape and design of the structural component, making it structurally undesirable [79]. Bonded repair is usually more reliable than bolted repair since bonding produces no holes and therefore reduces regional stress. Bonded repair contains a series of complex processes and has to be undertaken by well-trained technicians. In addition to the strict in-process control, postprocess nondestructive inspection (NDI) is necessary to guarantee the quality of the bonded joint. A general procedure of the repair activities for composite structures is described in Fig. 8.3.

There are many source documents containing information on maintenance, modification, and repair. One of the most complete maintenance documents in terms of instructions for damage disposition, inspection, and repair is the structural repair manual (SRM). It provides general airplane data, usual procedure, and repair materials for the repair of the specific type of aircraft. Service bulletins (SBs) are the documents issued by an original equipment manufacturer (OEM) that share modifications to previous maintenance instructions and include supplemental inspection, rework, and repair for a specific component. In some cases, there may be damage caused by unanticipated secondary loads. Service newsletters are issued by OEM to make both users and operators aware of any potential damage. Other documents pertinent to maintenance field are aircraft maintenance manuals (AMMs), component maintenance manuals

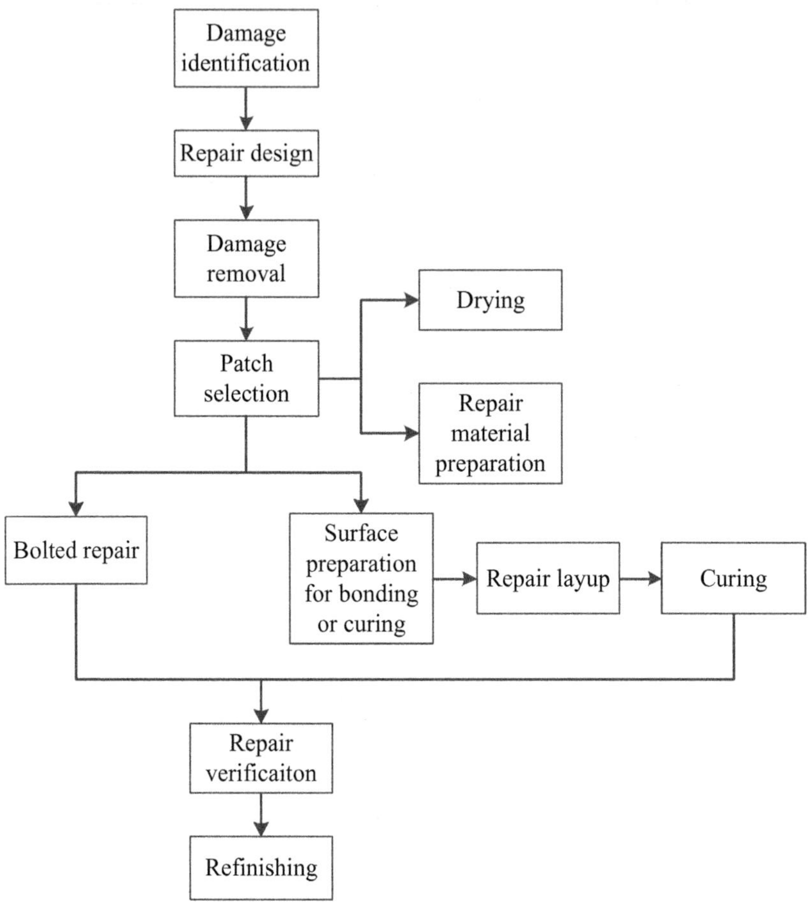

FIGURE 8.3 Repair flow chart.

(CMMs), and others. They are referred to as instructions to implement particular repair activities.

There have been significant research publications examining various repair techniques for composite structures, but little information is found concerning under what conditions these repair activities should be carried out, in other words, the repair threshold. Furthermore, not every damaged structure is repairable. One situation is that the damage is too severe to repair. The other situation is that considering influencing factors, such as time and cost, it is more convenient or economic to replace it by a new component instead of repair. Hereafter, a concept called "repair tolerance" is proposed to describe the repair thresholds and this chapter is devoted to the development of a probabilistic method to address the repair tolerance problem.

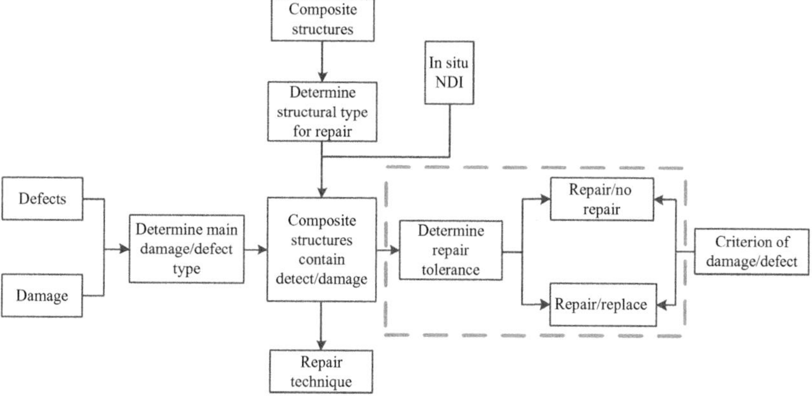

FIGURE 8.4 Content of repair tolerance research.

8.2 REPAIR TOLERANCE

The composite repair issue is a kind of system engineering that contains various parts, such as damage characteristics, structural properties, damage detection, repair techniques, and so forth, as displayed in Fig. 8.4.

The highlighted rectangular part is the focus of this study, in which repair tolerance is determined. The concept of repair tolerance was first proposed by Shaojie Chen [78]. It defined two thresholds—when to repair and when to replace—as visualized in Fig. 8.5. It is shown that repair tolerance is a subset of damage tolerance. Damage tolerance is one of the design requirements for structural design, which looks into the residual performance of a damaged structure and designs based on it. In comparison, repair tolerance is the measurement for structural reparability, which also studies the residual performance after damage based on the requirements for structural strength and stiffness. Besides, it should further combine many other factors, such as the repair technique, human factor, maintenance cost, and so forth, to determine the detailed policy.

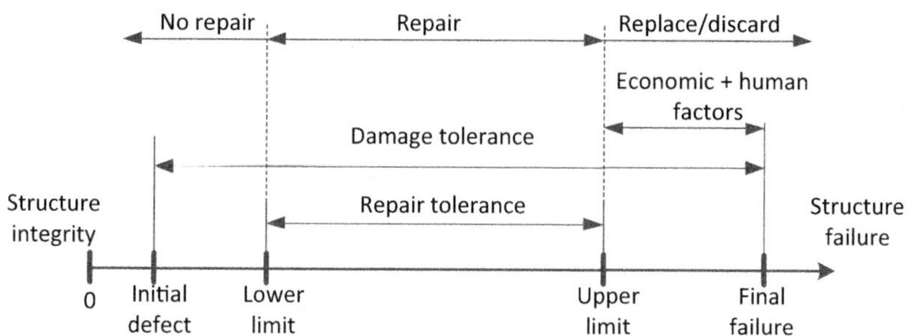

FIGURE 8.5 Repair tolerance concept.

Reparability is the degree of difficulty with which the damaged structure can be maintained or repaired in order to restore its strength or functionality according to specific requirements by referring to the structural repair policy. As a result, the determination of repair tolerance depends on the demonstration of the reparability which should consider various influencing factors, such as damage/defect mode, damage size and degree, repair condition, the qualification of repair personnel, economic effectiveness, and so forth. Theoretically, the parameter that the repair tolerance controls directly is the residual strength ratio (residual strength/designed strength). However, in practice, damage sizes are more intuitive and the relationship between the residual strength and the damage size can be obtained by theoretical calculations or laboratory experiments. For the sake of simplicity, damage size is used as the independent variable to describe repair tolerance.

Similar to the damage category of metallic airframes, composite structural damage can be classified as allowable damage, repairable damage, and unrepairable damage. Corresponding to the lower threshold and upper threshold in repair tolerance, the lower limit controls the threshold whether the damage is allowable or should be repaired, whereas the upper limit controls the threshold whether the damage is repairable or unrepairable, meaning a replacement is required. If the lower threshold is set too low, even minor damage will initiate a repair activity that may be unnecessary and costly; if the lower threshold is set too high, the structure may remain in a damaged condition for a long time, which may cause a threat to safety. Therefore, the lower threshold has a major influence on safety. As for the upper threshold, since the damaged structure is destined to go through certain maintenance work, the selection of a cost-effective maintenance task becomes the engineers' primary concern. Overall, dealing with uncertainty in the two thresholds has a significant impact on both safety and economy. In this chapter, the probabilistic approach developed in the previous chapter on inspection interval optimization is updated to quantify the repair tolerance from two aspects: the uncertainty in the lower threshold is measured by the risk of safety and the uncertainty in the upper threshold is assessed by the maintenance cost.

8.3 PROBABILISTIC METHOD

The probabilistic approach for the damage-tolerant composite structures is addressed with a computer simulation. Many sampling-based probabilistic methods have been proposed, such as first and second order reliability method, importance sampling method, advanced mean-based method, and Markov chain analysis [74,80]. These methods are either based on fast probability integration (FPI), which only works well for smooth performance functions or are limited by too many assumptions. The task here involves a series of discrete random variables describing the accidental damage and maintenance activities for composite airframes in-service. Monte Carlo simulation is considered as a viable

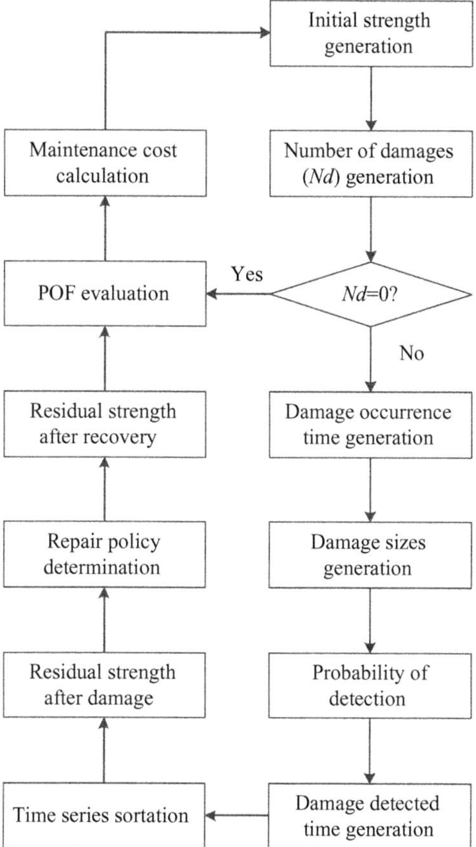

FIGURE 8.6 Simulation flowchart.

alternative that can easily handle complex scenarios and is proved to be a robust and flexible modeling approach [72].

The realistic conditions for composite structures incorporating accidental impact, strength reduction after impact, periodic inspection, and necessary repair or replacement in an operational life cycle, and so forth can be simulated in the procedure, as shown in Fig. 8.6. Because of the random behavior of damage, there exists uncertainty from one simulation to another. Therefore, the simulation should be conducted many times to tackle the scatter problem. Although time consuming, it can provide more accurate and practical results.

The sequence of the simulation procedure is illustrated as follows:

1. The first step is to generate the initial strength for the composite structure. A Gaussian PDF is often used for strength analysis and product quality control [75]. It follows that the strength scatter introduced in the manufacturing process is described by a Gaussian PDF.

2. The damages occurring in a composite structure in operational life cycle are a series of discrete and rare events, and are best described by a Poisson distribution.
3. If the generated number of damage (*Nd*) is 0, that is, no damage has occurred, and except for scheduled inspection, no maintenance activities are required. Then the probability of failure (POF) and maintenance cost can be evaluated, which will be discussed at the end of this section.
4. If $Nd > 0$, the damage occurrence time is generated. Since impact damages caused by runway debris, hail, human mishandling, and so forth are highly random and accidental, a uniform distribution generator is used to describe the scattered damages that may possibly occur at any time throughout the service life.
5. After the generation of damage occurrence time, damage sizes are generated. The distribution model can be derived based on damage records statistics from real operational aircraft.
6. The inspection efficiency is described by the probability of detection (POD). It is a time-consuming effort to obtain an exact relationship of POD against the damage size. Multiple influencing factors should be taken into account, such as colors of the composite panel, cleanness, brightness, inspection angles, personal eyesight, professional skills, and so forth [68]. Generally there are three inspection levels: general visual inspection (GVI), detailed inspection (DET/DI) and special detailed inspection (SDI). For structures in different locations, an appropriate inspection level should be selected and the corresponding POD can be obtained from a certain probability function.
7. Since POD is introduced, the time t to detect damage may be delayed to the subsequent inspections and is expressed as:

$$t = T \times n,$$

where $\boldsymbol{T}$ is the predetermined inspection interval and $\boldsymbol{n}$ is the number of times to detect the damage, which can be generated by a geometric distribution.
8. Assuming that the inspection time and repair time are negligible, the damage occurrence time and damage detection time should be ordered in sequence to facilitate the description of residual strength variation.
9. The damage size should be converted to the reduction of residual strength. The relationship of residual strength against damage size for a particular composite structure can be obtained by experiment or theoretical calculation.
10. After damage, the following inspection offers a window for damage disposition that whether it should be repaired and what repair process should be taken, provided that damage is detected. The determination of the repair policy has a direct effect on the probability of failure (POF) and the maintenance cost, which is a series of proactive actions after each inspection that can be controlled. If the damage has not reached the critical level, no

repair is needed or simply basic repair is applied to protect and decorate the surface. If the damage is beyond the critical level, two types of repair are usually initiated: temporary repair and permanent repair. Specifically, for laminates and sandwich panels, there are three basic approaches. First is patch repair, which is quick and simple but without consideration of thickness and weight increases. This belongs to temporary repair. The other two are scarf repair and step sanded repair, which can provide a straighter and stronger load path but requires time and high skill [79]. They are permanent repairs. If the damage is too severe, replacement may be more efficient than repair.

11. According to different repair techniques, the residual strength will be recovered to different levels. If the structure is repaired, a recovery efficiency coefficient will be appended by a uniform distribution within a certain range. If the structure is replaced by a new one, the generation of the strength will be described by a Gaussian PDF.
12. The probability of failure (POF) is calculated by the following formula:

$$\text{POF} = 1 - \prod_{i=1}^{N} \left[1 - P_f\left(S_i, t_i\right)\right] \tag{8.1}$$

where $\boldsymbol{t_i}$ is the $\boldsymbol{i}$th time interval between ($\boldsymbol{i}$–1)th and $\boldsymbol{i}$th activity (0 means the initial service time), $\boldsymbol{S_i}$ is the $\boldsymbol{i}$th residual strength between ($\boldsymbol{i}$–1)th and $\boldsymbol{i}$th activity, $\boldsymbol{N}$ is the number of damages occurred in one life cycle, and $\boldsymbol{P_f}(\cdot)$ is the probability of failure for each interval with constant residual strength.

Failure occurs when the applied load exceeds the residual strength. Each time interval throughout the life cycle with constant residual strength is assumed in series connection. The cumulative distribution function (CDF) of the maximum load per t_i is expressed as:

$$F_l\left(S_i, t_i\right) = e^{-H(S_i)t_i} \tag{8.2}$$

where $\boldsymbol{H}(\boldsymbol{x})$ is the frequency of the event exceeding the level $\boldsymbol{x}$, which is described by different load exceedance curves after load cases are specified. A detailed illustration has been explained in the previous chapter.

13. The last step of the simulation cycle is to calculate the total maintenance cost. The total maintenance cost can be expressed as:

$$C_{\text{total}} = C_{\text{inspection}} + C_{\text{repair}} + C_{\text{replace}} + C_{\text{other}} \tag{8.3}$$

where $\boldsymbol{C}_{\text{total}}$ is the total maintenance cost; $\boldsymbol{C}_{\text{inspection}}$ is the costs incurred by scheduled inspection including the labor and equipment cost; $\boldsymbol{C}_{\text{repair}}$ is the repair cost including labor, material, and equipment costs; $\boldsymbol{C}_{\text{replace}}$ is the replacement costs including labor, equipment, and spare parts; and the last

C_{other} denotes any other cost that may be caused by unscheduled maintenance, flight delay, or other operational problems.

Overall, an operational life cycle of a composite structure is simulated by the aforementioned 13 steps. In order to address the uncertainty, the simulation should be repeated with a large sample to obtain the mean value of the POF and the maintenance cost.

8.4 CASE STUDY

According to statistics from the previous survey in an airline maintenance department, impact damage caused by natural object and human mishandling is the most frequent damage type. Therefore, impact damage resulting in dent or delamination was assumed to be the only damage type. A GFRP composite wing panel was selected. Statistical input data and related assumptions are listed later.

8.4.1 Load Case

For civil aircraft structures, gust load is mainly considered as the critical load case [73]:

$$P_{rat} = P(\varepsilon > \varepsilon_{LL}) = 2\times10^{-5}/\text{FH};$$

$$P_{rat} = P(\varepsilon > \varepsilon_{UL}) = 1\times10^{-9}/\text{FH}$$

where ε is the actual load, ε_{LL} is the limit load, and ε_{UL} is the ultimate load, $\varepsilon_{UL} = 1.5\varepsilon_{LL}$.

Since the probability value under different exceeding conditions changes significantly by the power of 10, a log-linear model is used to describe the load occurrence probability. The load exceedance curve is shown in Fig. 8.7.

$$\lg(P_{rat}) = -8.602\times\left(\frac{\varepsilon}{\varepsilon_{LL}}\right)+8.903 \tag{8.4}$$

8.4.2 Average Damage Per Life Cycle (Nd)

Information on dent and delamination was extracted from the maintenance records from 2002 to 2012 about wing structures of a certain aircraft fleet. In total 19 occurrences of impact damage were recorded in four GFPR panels from 12 aircrafts. The design life of the aircraft is 150,000 flight hours (50 years) and the composite panel is assumed to have the same design life. Thus the average number of damage per operational life cycle is approximately 2.

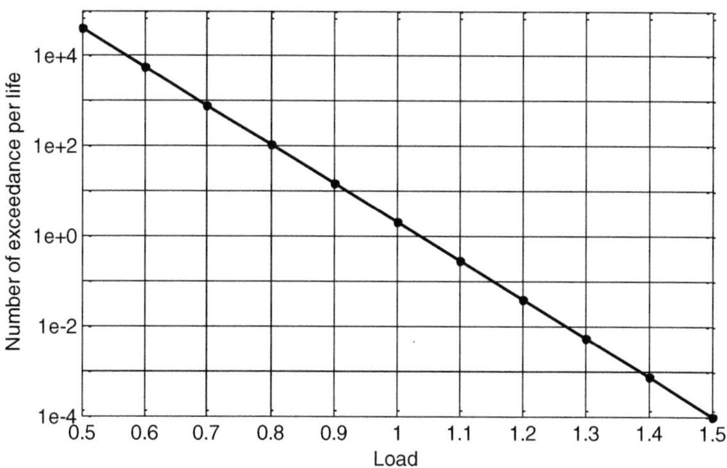

FIGURE 8.7 Load exceedance curve.

8.4.3 Damage Size Distribution

Statistical analysis was performed to obtain the probability distribution function (PDF) of the damage size. Four most likely PDFs were tested and their goodness-of-fit are shown in Fig. 8.8.

It is seen from the test that the Weibull distribution function is the best PDF with the highest correlation coefficient of 0.995. The histogram of the damage diameter distribution is plotted in Fig. 8.9. Note that all the damage sizes in this chapter are in inches.

The Weibull expression of this distribution is given by:

$$f(d;\alpha,\beta)=\frac{\beta}{\alpha^{\beta}}d^{\beta-1}e^{-\left(\frac{d}{\alpha}\right)^{\beta}},\quad d\geq 0,\quad \alpha>0,\quad \beta>0$$
$$\alpha=3.486,\quad \beta=1.950$$

Therefore, a Weibull generator is used for impact damage sizes.

8.4.4 Probability of Detection (POD)

From previous analysis, two cumulative probability functions are used to describe the POD against damage size [69,74], the cumulative Weibull distribution function:

$$F(x)=1-e^{-\left(\frac{x}{\beta}\right)^{\alpha}}$$

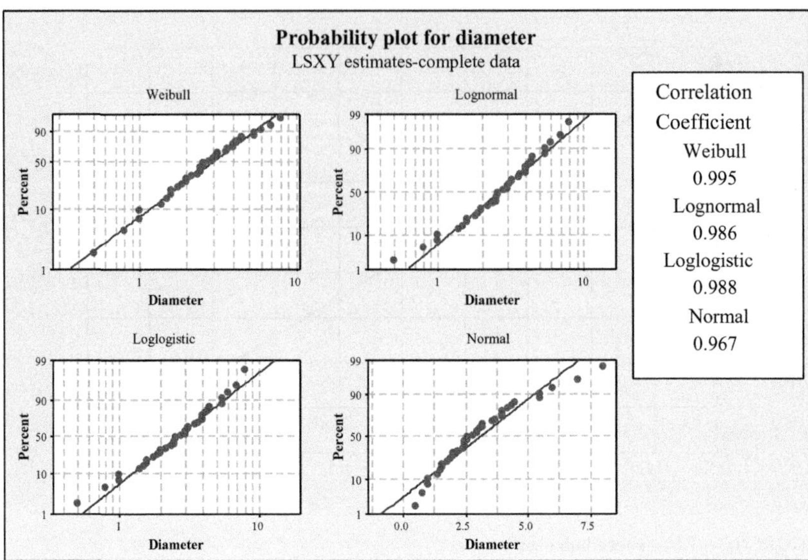

FIGURE 8.8 Goodness-of-fit test for damage size.

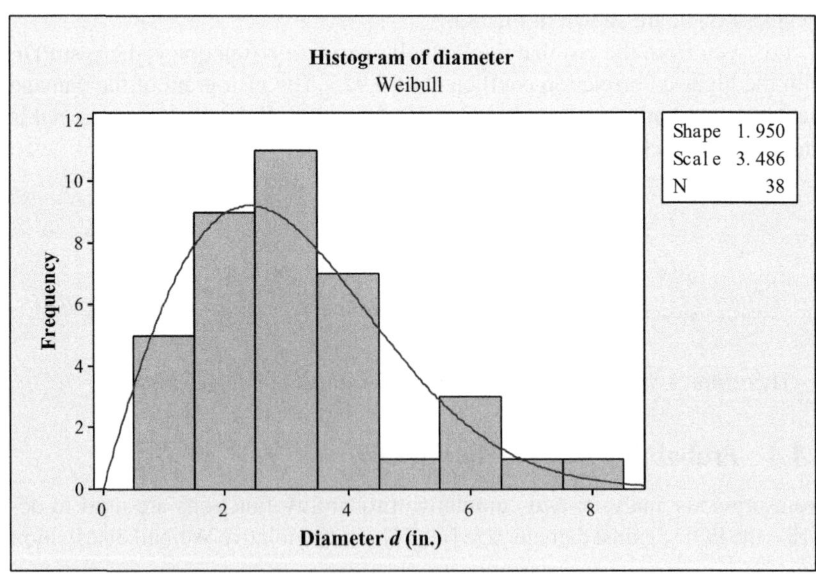

FIGURE 8.9 Damage diameter distribution.

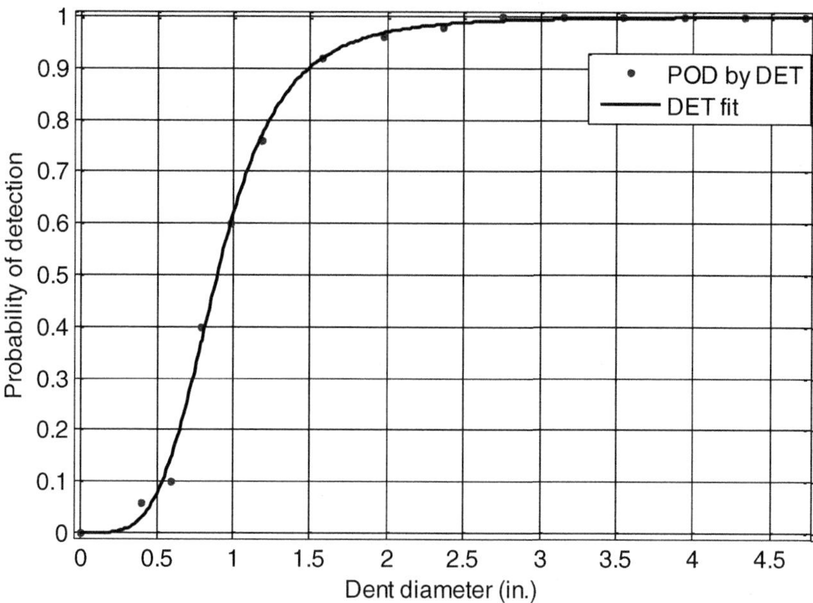

FIGURE 8.10 **POD versus damage diameter (in.).**

and, the cumulative log-normal distribution function:

$$F(x) = \frac{e^{\alpha+\beta\ln(x)}}{1+e^{\alpha+\beta\ln(x)}}$$

For the selected composite panel, detailed inspection (DET/DI) is considered and log-normal model proves to fit the data well [70] with a higher regression value compared to Weibull, and the expression is given here (Fig. 8.10):

$$\text{POD\_DET}(d) = \frac{e^{\alpha+\beta\ln(d)}}{1+e^{\alpha+\beta\ln(d)}}, \quad \alpha = 0.4877, \quad \beta = 4.294 \tag{8.5}$$

8.4.5 Inspection Schedule

The selection of the inspection interval is based on the real maintenance schedule of the airline obtained from the Maintenance Review Board Report (MRBR). The inspection interval for the composite panel is a "4C" check, which is 16,000 flight hours by DET. Since the service life is 150,000 flight hours, the interval is approximated to 15,000 flight hours (10% of the operational life cycle) to facilitate the calculation in the numerical example.

8.4.6 Residual Strength Reduction and Recovery

The relative residual strength of the GFRP composite panel as a function of damage size for impact damage is approximated by the following function:

$$\mathrm{RS}(d)=\begin{cases} 1-kd, & 0<d<1,\quad k=0.05 \\ A+(C-A)e^{\left(\frac{d}{G}\right)}, & d\geq 1,\quad A=0.46,\quad C=2.3189,\quad G=0.75 \end{cases} \tag{8.6}$$

where A is the residual strength asymptote; C is the intercept; G and k are the slopes for the segmented curve.

The recovery of the residual strength depends on different repair policies, which will be discussed in the Section 8.4.7. Generally, the recovery efficiency coefficient is described by a uniform distribution within the range of 0.85–0.95.

8.4.7 Repair Policy

A simplified model is assumed here for the repair tolerance concept. The lower threshold and upper threshold are used as two decision variables that divide the damage degree into three intervals, as shown in Fig. 8.5.

1. If the damage size is less than the lower threshold, the structure is left as it is;
2. If the damage size is between the lower threshold and upper threshold, repair work is initiated and considered as permanent repair that does not require a change in the inspection schedule;
3. If the damage size is larger than the upper threshold, the damaged structure is replaced by a new spare.

8.4.8 Factor of Safety

The scatter of the new strength is described by a Gaussian PDF with a coefficient of variation of 5%. The mean value of initial strength is calculated as:

$$\mathrm{RS}=f\times f_1=1.5\times 1.4=2.1,$$

where f = 1.5 is the factor of safety and f_1 = 1.4 is the additional margin of safety. Environmental factors, such as temperature, moisture, and ultra-violet may induce slow degradation of composites. Also, adjacent damage may result in additional strength loss. For simplicity, these factors are not considered in this study but will be investigated in the future.

8.4.9 Probability of Failure (POF)

The operational life cycle is set to 1 unit; accordingly the inspection interval T = 0.1. As mentioned before, the lower threshold has a direct influence on structural safety, which is described by the probability of failure (POF). The upper threshold is fixed at 5 in. as an example. Note that the upper threshold can be

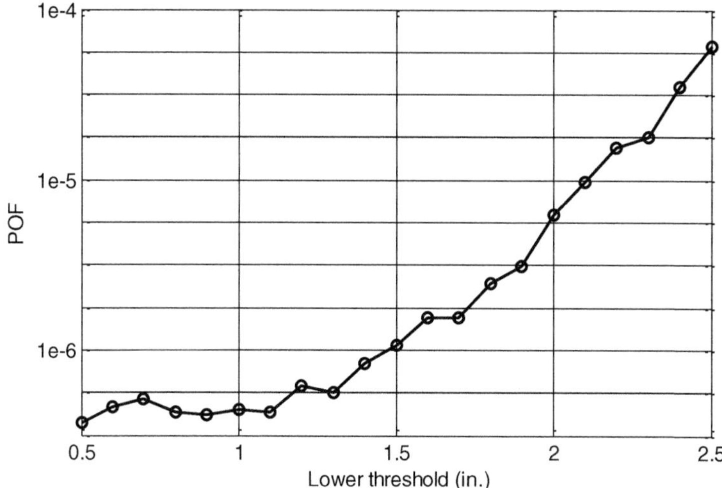

FIGURE 8.11 **POF versus lower threshold.**

fixed at any value as long as reasonable because it has little effect on structural safety. After taking 1000 samples, the average probability of failure (POF) in relation to the lower threshold is obtained and shown in Fig. 8.11.

8.4.10 Maintenance Cost

The maintenance costs consist of four parts: inspection cost, repair cost, replacement cost, and risk cost. The risk cost is induced by any unpredicted failure when the POF is above a certain level. While of the lower threshold affecting safety, the upper threshold plays an important role in the maintenance cost. As discussed in the repair tolerance section, various influencing factors should be taken into account, such as damage severity, the repair capability (human and equipment), spare parts management, repair duration, and so forth. Thus, an integrated exponential expression in relation to the damage size is assumed, based on airline operational experience considering multiple factors:

$$C_{\text{repair}} = C_{\text{basic}} + C_1 e^{C_2 d_i} \tag{8.7}$$

where C_{basic} is the basic cost for every repair, C_1 and C_2 are coefficients according to practical maintenance conditions balancing capability, time, spare parts, and so forth, d_i is the ith damage size.

In terms of the lower threshold, if it is set too large, the POF may be lower than the required level and thereby, resulting in a larger possibility to fail, unscheduled maintenance will then be initiated causing more labor work and flight

TABLE 8.1 Cost for Each Maintenance Task

$C_{inspection}$	—	100
C_{repair}	C_{basic}	600
	C_1	100
	C_2	0.6
$C_{replacement}$	—	4,000
C_{risk}	If POF > 1e-3	10,000

delays. Therefore, a cost of risk is introduced to incorporate the lower threshold into the maintenance cost model.

Values for each cost type are assumed based on airline operational experience shown in Table 8.1. Note that the unit is omitted.

The range of the lower threshold variable is set to 0.5–2.5 (in.) and of the upper threshold variable is set to 4.0–7.5 (in.), 0.1 (in.) a step value. After taking 1000 samples, the average maintenance cost against the two thresholds is obtained and shown in Fig. 8.12.

8.5 RESULTS AND DISCUSSION

It is seen in Fig. 8.11 that, generally, the probability of failure (POF) follows a monotone trend with the lower threshold except for the fluctuation within 0.5–1 in. This is because the residual strength decreases slightly in this range,

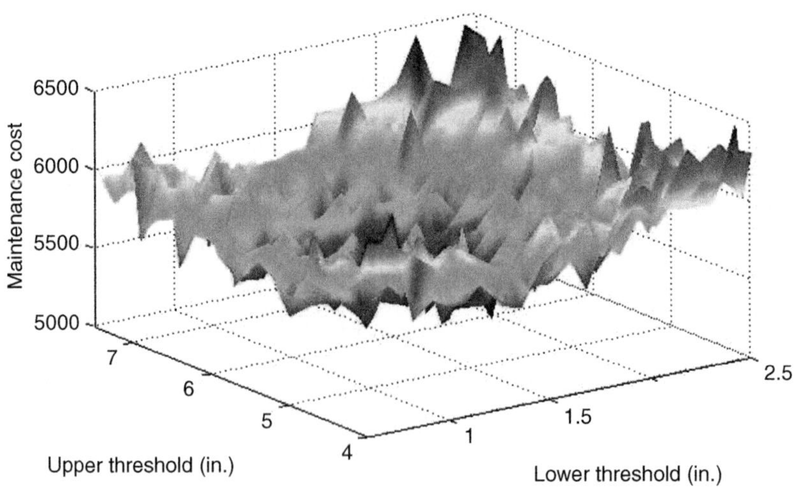

FIGURE 8.12 Maintenance cost versus repair tolerance (lower and upper thresholds).

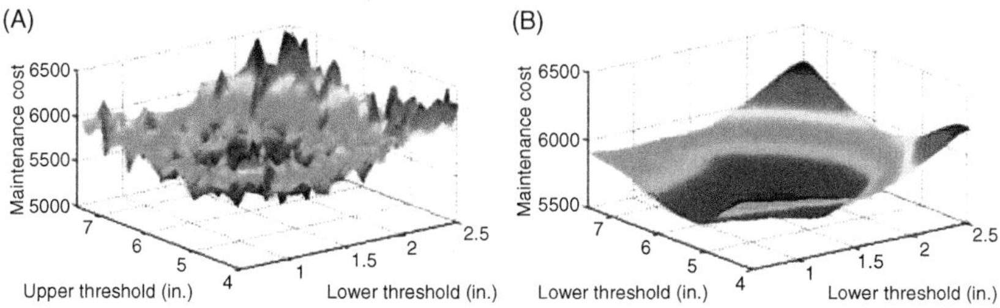

FIGURE 8.13 Maintenance cost wavelet denoising.

causing little effect on the load-bearing capacity of the whole structure. By assigning the lower threshold a risk of safety, which is described by a POF value, airline engineers can determine the exact lower threshold value by assessing the required reliability of the composite structure in different situations. For instance, if a requirement for the structural probability of failure is no higher than 10^{-5}, the maximum lower threshold should be 2.1 in., which means any damage within 2.1 in. can be kept as it is.

Due to the large scatter in each simulation, a large number of samples are required, which is time-consuming. As reflected in Fig. 8.12, the surface of maintenance cost is too variable to determine any particular trend and the best value. Taking the maintenance data as signals, a denoising process was implemented by wavelet analysis, which is capable of revealing aspects of data like trends, breakdown points, self-similarity, and so forth [81]. There are generally two steps for denoising. First, a noisy signal is decomposed, and then the other half of the process is reconstructed. A comparison of the maintenance cost before and after wavelet denoising is shown in Fig. 8.13.

A much smoother surface is obtained on the right-hand graph that shows a relative clear trend of the cost variation while keeping an acceptable accuracy level. Therefore, wavelet denoising provides a good balance between simulation time and accuracy. Further, a contour plot representation of Fig. 8.13 (B) is shown in Fig. 8.14.

The minimum maintenance cost in Fig. 8.14 is highlighted by "+" symbol, approximately 5540. Accordingly, the optimized lower threshold and upper threshold in repair tolerance are 1.6 and 5.5 in. respectively for the selected composite structure.

The optimization of the lower and upper threshold depends largely on the cost ratio of different maintenance tasks as shown in Table 8.1. Expenditures on scheduled maintenance activities, such as inspection, repair, and replacement vary a lot according to different labor cost, equipment cost, and spare part management capability of airlines in different regions. So, does the risk cost induced by unscheduled maintenance and flight delays in case of a high probability of failure. The cost ratio variation of maintenance tasks can result in different optimized repair thresholds.

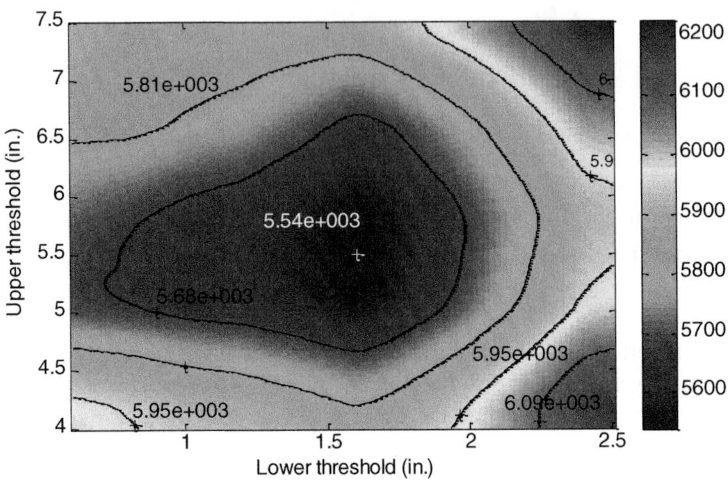

FIGURE 8.14 **Contour map of maintenance cost (optimization of maintenance cost).**

Other influencing factors are listed as follows:

1. The design of the composite structure, of which the residual strength decreases differently against the damage size;
2. Applied load cases, which correspond to different probability of failure;
3. Inspection schedule, which is undergoing a gradual shift from time-based inspection to condition-based inspection with the development of structural health monitoring (SHM) techniques;
4. Operational environment and personal qualification, which has a direct relation with the occurrence frequency of the impact damage caused by natural accident and human mishandling.

Overall, the repair tolerance is not simply a design index like damage tolerance. The determination of the repair tolerance should also rely on practical operation as well as maintenance situation. It is an integrated index throughout the life cycle of the aircraft structure.

8.6 CONCLUSIONS

In this chapter, a concept of repair tolerance was proposed by defining two thresholds for damaged composite airframes. The lower threshold denotes whether to repair and the upper threshold refers to whether to repair or replace. A probabilistic approach was applied to address uncertainty problems in repair tolerance through Monte Carlo (MC) simulation. The uncertainty in the lower threshold was assessed by a risk of safety whereas the uncertainty in the upper threshold was evaluated by considering economic and human factors. Two optimal thresholds in repair tolerance were derived based on large sample iteration in order to minimize the total maintenance cost. A typical composite structure from an in-service aircraft type was selected as a numerical example to demonstrate the effectiveness of the methodology.

Though this is only an idealized model based on damage information partly based on assumption and partly obtained from real operation, the probabilistic method is quite flexible and can potentially incorporate more practical modules in the future, for example, a more detailed repair policy and maintenance cost module can be developed. Also, this method can be expanded to minimize the life-cycle cost (LCC) by taking design and manufacturing expenses into consideration.

Chapter 9

Structural Health Monitoring and Influence on Current Maintenance

Chapter Outline

9.1 STRUCTURAL HEALTH MONITORING TECHNOLOGY

Structural health monitoring (SHM) is defined as the process of implementing a damage identification strategy for aerospace, civil, and mechanical engineering infrastructure [82,83]. The damage inspection and identification is based on a wide variety of highly effective local nondestructive evaluation devices. Superficially from the definition, SHM is simply an implementation of an advanced monitoring technique by a variety of sensors. But the final objective of applying SHM is to realize a complete condition-based maintenance (CBM), in which the monitoring process is autonomous and continuous. SHM is thus able to liberate the complicated inspection work and reduce the cost involved with aircraft downtime during maintenance. In the long term, the ultimate goal of SHM is to reduce the requirement for overdesign of aircraft structures and increase performance by incorporating reliable SHM systems at the design stage. Below the surface of installing advanced sensors or even creating smart structures, there is a considerable comprehensive system involving multidisciplinary technologies.

The SHM incorporates various modules, such as sensors deployment and installation, data transition and storage, data cleansing and fusion or simply data management, damage diagnosis and prognosis, maintenance decision making, and so forth, as shown in Fig. 9.1.

Reliability Based Aircraft Maintenance Optimization and Applications
http://dx.doi.org/10.1016/B978-0-12-812668-4.00009-5

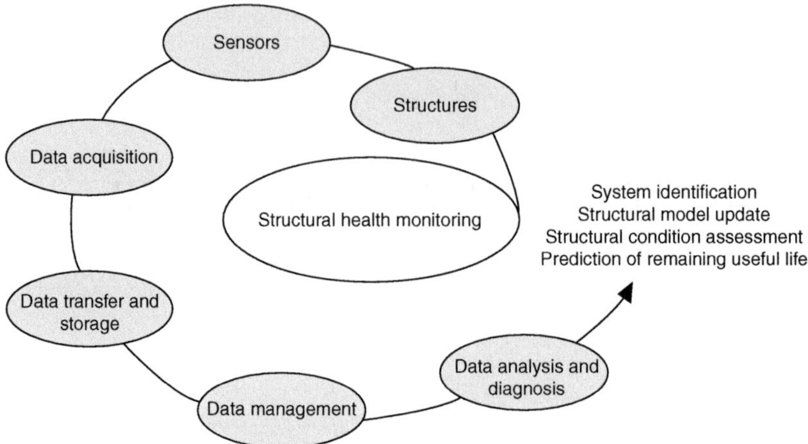

FIGURE 9.1 Structure health monitoring (SHM) modules.

9.2 SHM APPLICATIONS IN AIRCRAFT

A wide range of sensors, either attached or embedded, have been studied and developed; typical examples are strain gauges, accelerometers, fiber-optic sensors, acoustic sensors (passive and active), and so forth. After the collection of data via various sensors, a complete SHM system should be capable of performing diagnosis and prognosis functions in order to assist operational decision making, such as maintenance and logistic support. In commercial aircraft, the SHM application generally cares for two critical aspects, that is, operation loads monitoring and impact damage detection. Operational load monitoring is used to support the assessment of structural fatigue life by measuring the local stresses either directly or indirectly [84,85]. Electrical strain gauges can be considered as the most mature tool for load monitoring. Operational load monitoring is applied widely to different types of aircraft for evaluation of accumulated fatigue damage and remaining structural life. However, it cannot provide direct damage information, such as metallic corrosion and delamination of composites. Therefore, the other important aspect is required: damage detection that allows the direct measurement of potential damage onto or into aircraft structural components. Ultrasonic/acoustic nondestructive technology is the current proven method for damage detection [86]. Alternatively, fiber optic strain sensors are increasingly used with the advantage of multifunctional measurement capability in an integrated system, which can significantly simplify the complexity of the SHM system.

For composite structures, eight scenarios of potential SHM applications were identified, associated with specialists from stress, design, maintenance, and repair [87]:

1. Impact damage
2. Stringer/skin interface failure

3. Debonding of carbon fiber reinforced plastics (CFRP) cobonded parts
4. Core/skin sheet debonding in sandwich structures
5. Delamination of CFRP-skin layers
6. Damage of honeycomb structure
7. Detection of missing rivet heads in CFRP structures
8. Detection of loads and stress/strain distribution in CFRP structures

Recent researches began to recognize that the SHM problem is fundamentally one of statistical pattern recognitions [82], which involves four processes: (1) operational evaluation, (2) data acquisition, fusion, and cleansing, (3) feature extraction and information condensation, and (4) statistical model development for feature discrimination. From a perspective of CBM, it mainly consists of three key steps: (1) data acquisition, to obtain and store data relevant to structural health, (2) data processing, to analyze signals or data collected for well interpretation, and (3) maintenance decision-making, to recommend effective maintenance policies [88].

A general architecture for a complete SHM system and its operation within an aircraft maintenance program is described in Fig. 9.2.

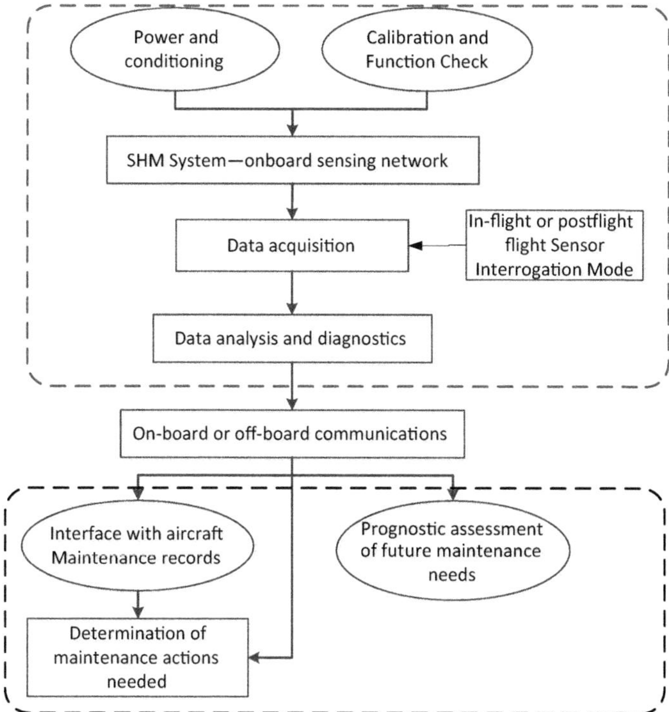

FIGURE 9.2 SHM system operation within aircraft maintenance.

Many researches and publications on a wide variety of SHM techniques belong to the upper portion of the diagram, which is the on-board monitoring system or complex data processing and analysis methods in ground-based systems. The final objective of SHM is to facilitate maintenance activities and further, to improve structural design. This chapter focuses on the other portion of the diagram and investigates the problem: how to integrate SHM tasks into the current maintenance program?

9.3 INFLUENCE OF SHM ON CURRENT MAINTENANCE

SHM is receiving increasing attention in some official documents. MSG-3 is one of the current maintenance practices that combine more than 40 years experience from aircraft manufacturers, airlines, and regulatory authorities to determine efficient maintenance tasks. In order to keep it up-to-date to the new technologies, regulations, and maintenance processes, it was in the revision 2009.1 that the MSG-3 document included SHM and scheduled SHM (S-SHM) concepts for the first time but has not designed any explicit logic decision process [8].

Four years later, in 2013, an SAE technical report named ARP 6461 was published by G11SHM, Structural Health Monitoring and Mgmt (AISC). This document provided key definitions, guidelines, and examples for civil aerospace airframe structural applications in the development, validation, verification, and certification of SHM systems [89]. The release of the document fulfills at least two purposes:

1. Due to the diversity in defining and classifying SHM, ARP 6461 can serve as a standard to facilitate the worldwide standardization and harmonization of understanding about SHM.
2. Due to the immaturity of SHM, ARP 6461 can provide information on structural maintenance practices and provide guidance on how SHM can be incorporated with or as modifications to current maintenance and airworthiness documents. Relevant definitions in MSG-3 and ARP6461 used in the thesis are listed as follows:
   a. *SHM*: The concept of checking or watching a specific structural item, detail, installation, or assembly using on board mechanical, optical, or electronic devices specifically designed for the application used (MSG-3); The process of acquiring and analyzing data from on-board sensors to determine the health of a structure (ARP6461).
   b. *Scheduled SHM (S-SHM)*: The act to use/run/read out a SHM device at an interval set at a fixed schedule. (MSG-3)
   c. *Automated SHM (A-SHM)*: Automated SHM is any SHM technology that does not have a predetermined interval at which maintenance action must takes places, but instead relies on the system to inform maintenance personnel that action must take place. (MSG-3)

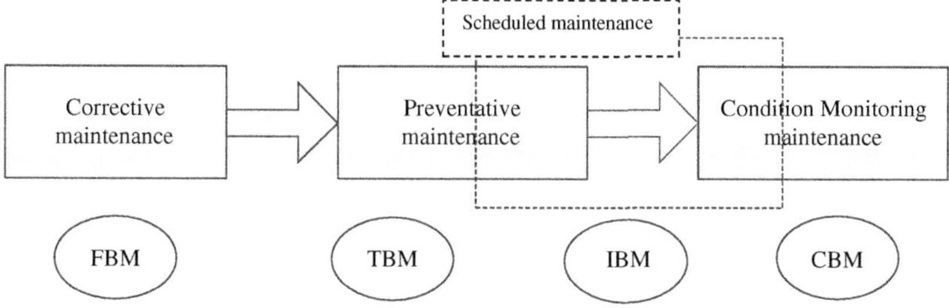

FIGURE 9.3 Evolution of maintenance strategies and policies.

Maintenance is a key activity in aircraft operations to ensure its continuous safe and effective operation. An efficient maintenance program can save cost over the aircraft operational life. Maintenance strategies have evolved from corrective maintenance to preventative maintenance. In the early stage, corrective maintenance activities are initiated after the occurrence of a fault or failure, which is fault-based maintenance (FBM) [90]. With increasing knowledge complexity of the system, preventative maintenance is carried out at predetermined intervals to detect and repair damage. Early preventative maintenance is also called time-based maintenance (TBM), in which components are replaced at fixed times at the sacrifice of residual useful life. Gradually, inspection activities are used to replace previous hard-time replacements, which is inspection-based maintenance (IBM) and it is the kind of scheduled maintenance carried out nowadays. More precisely, repair activities are performed based on the condition of the structural component by periodic inspections. With the development of SHM technology, the health condition of a system or product can be monitored more frequently with lower cost or even continuously, opening the way to condition-based maintenance (CBM) [91]. A general evolution of maintenance strategies and policies is shown in Fig. 9.3.

Scheduled maintenance sits between preventative maintenance and condition monitoring maintenance. The maintenance intervals and tasks are developed based on the document guidance called Maintenance Steering Group (MSG-3), which is accepted by regulatory authorities, operators, and manufacturers [8]. It is widely adopted in the commercial aviation industry for the development of minimum required scheduled maintenance procedures for both initial and continued airworthiness [92]. These maintenance intervals and tasks are determined in the design and manufacturing stage and, if necessary, updated during service based on customer feedback. It is typically one of the representations of concurrent engineering, which emphasizes the consideration of possible factors throughout aircraft life cycle. However, MSG-3 is facing shortcomings with the development of next generation aircraft, due to the use of advanced composite structures and the application of a variety of SHM systems. These developments exert a strong motivation to incorporate new concepts into the MSG with a shift from preventive maintenance to condition monitoring maintenance. SHM, which refers to the process of structural damage identification by acquiring and analyzing data from on-board sensors so that the health state of the structure can be monitored on a continuous basis [89]. SHM is playing an important role in facilitating the transformation from time and inspection-based maintenance to Condition-Based Maintenance (CBM). However, barriers from technical, systemic, and knowledge aspect prevent its effective application in commercial aircraft operations [93]. In the interim, an investigation into a combination of SHM and the current MSG-3 procedure is worth considering.

9.4 INTEGRATION OF SHM WITH MSG-3 ANALYSIS

As an overarching plan of incorporating SHM into current maintenance practices to guide both short-term and long-term maintenance activities is significantly required, documents were released or updated to address this problem. Specifically, the Issue Paper 105 was published to modify the MSG-3 logic to allow consideration of SHM applications for new design [94, 95]. SAE ARP 6461, as mentioned previously, was released to standardize and harmonize world understanding about SHM [89]. MSG-3 revision 2009.1 points out that emerging technology, such as SHM may be an option to examine AD, ED, and/or FD if demonstrated to be applicable and effective [8]. Considering different maturity levels for various structural health monitoring systems, a flexible integration of structural health monitoring into MSG-3 logic procedure is proposed and shown in Fig. 9.4.

Herein, two terms are specified again to facilitate understanding. Scheduled-SHM (S-SHM) refers to any SHM system that must be interrogated by maintenance personnel in order to function at an interval set at fixed schedule. Automated SHM (A-SHM) is any SHM technology that does not have a predetermined interval at which maintenance action must take place, but instead relies on the diagnostic or prognostic information provided by the SHM system as triggers.

As shown in Fig. 9.4, after the determination of all SSIs' initial maintenance tasks, feasibility and applicability analysis of available SHM systems is performed. If no SHM system is suitable, the original scheduled maintenance is applied, which is grouped as scenario A. If an onboard SHM system can monitor structural health states continuously and give feedback on any abnormal information whenever abnormality occurs, A-SHM is fully applicable and thus a complete CBM can be achieved. Although condition-based maintenance (CBM) also includes the maintenance based on periodical inspections, in this study CBM only refers to the maintenance based on real-time health monitoring. Maintenance tasks through A-SHM are grouped as scenario D. A myriad of combinations of scheduled maintenance and SHM exist between scenario A and D. Herein, two typical combinations are presented in scenario B and C according to different deployments of sensor networks. Overall, the updated structural maintenance tasks can be roughly categorized as four scenarios: A. Scheduled Maintenance, B. Scheduled SHM, C. Scheduled CBM, D. CBM, reflecting three levels of SHM integration.

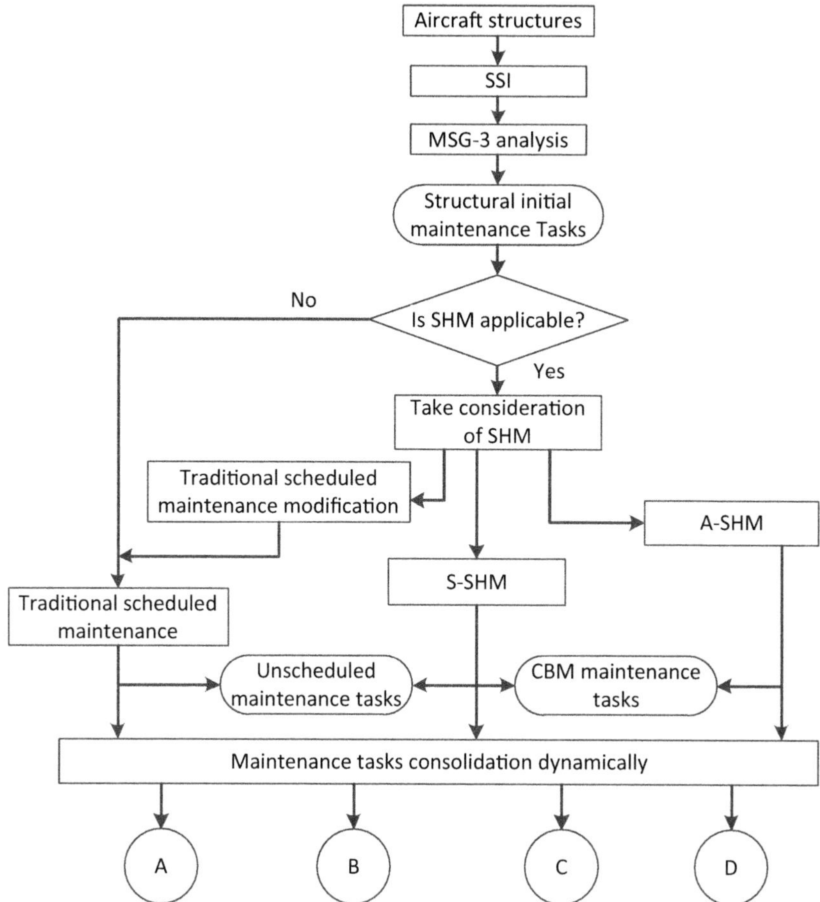

FIGURE 9.4 MSG-3 logic diagram considering SHM.

A. Scheduled Maintenance

Scheduled maintenance is performed at predetermined intervals to address damage remaining undetected in normal operations. During scheduled maintenance, aircraft usually stays in a hangar and undergoes intensive inspection and necessary repair work. Nondestructive inspection (NDI) is often performed, especially to inaccessible areas. However, disassembling and reassembling related structural components for inspection are inevitably required. Though time consuming, the detailed maintenance activities ensure aircraft safe operation until the next maintenance cycle. A typical logic procedure for maintenance of composite panels is shown in Fig. 9.5.

There are two critical parameters in scheduled maintenance affecting both safety and maintenance cost: the inspection interval and the repair threshold

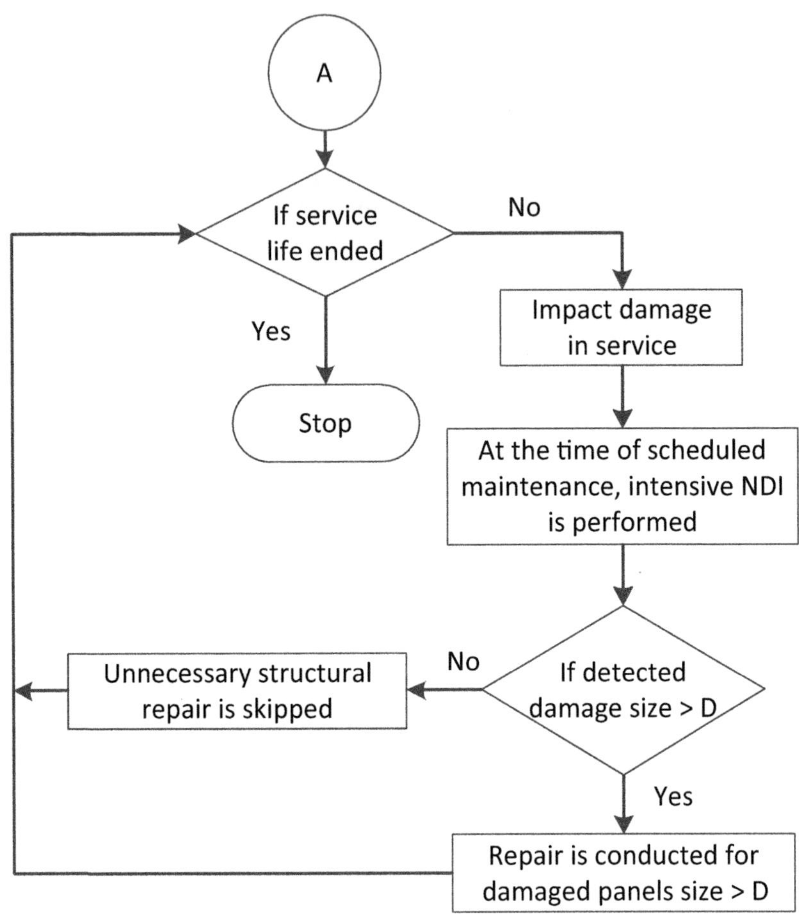

FIGURE 9.5 Scheduled maintenance procedure.

(denoted as D). With the integration of SHM in the following scenarios, these two parameters will be adjusted dynamically.

B. Scheduled SHM

Only a small fraction of structures undergo repair at each scheduled maintenance cycle, while every piece of SSI needs to be inspected to preclude any detrimental damages. Most NDI techniques for detecting damages, such as delamination in composites are labor intensive and costly.

In this scenario, an onboard SHM sensor system is implemented while the data collection and analysis system is ground based. The benefit is that the SHM system can detect damage without tearing down components and therefore

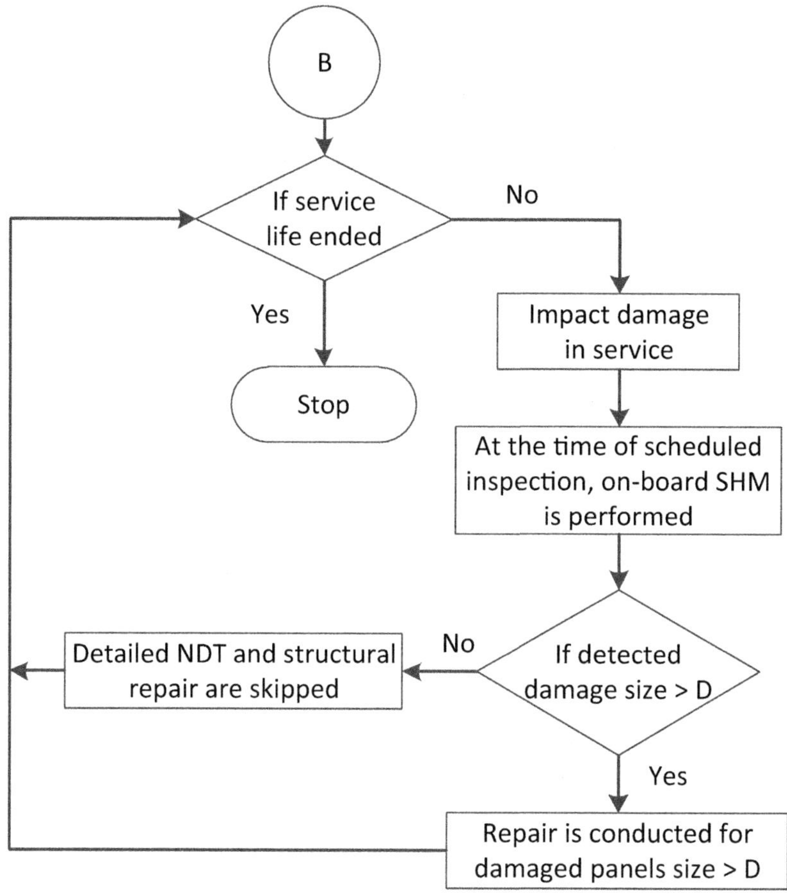

FIGURE 9.6 Scheduled SHM maintenance procedure.

intrusive inspections can be spared. It is noted that the inspection interval and repair threshold remain the same as those in scheduled maintenance, that is, the inspection of onboard SHM data is performed at every scheduled maintenance cycle. If damage exceeding the threshold is detected, repair activities are immediately initiated. Thus, the scheduled SHM can be seen as an updated version of the original scheduled maintenance. A logic procedure similar to scenario A is shown in Fig. 9.6.

Though labor saving, the damage detection by the SHM system may not be as reliable as the traditional intrusive inspection considering the immaturity of SHM technology and any secondary damage to SHM sensors. As a result, this maintenance scenario may lead to a lower safety level compared with scheduled maintenance.

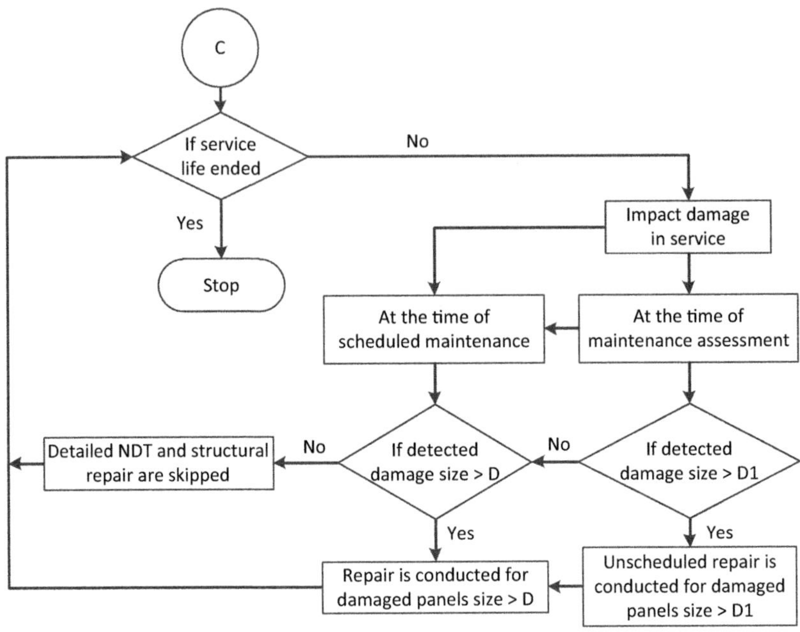

FIGURE 9.7 Scheduled CBM maintenance procedure.

C. Scheduled CBM

The alternative combination is designed to be a scheduled CBM maintenance philosophy. With increasing maturity of SHM, the data acquisition part is also placed in situ with the sensor network. The electronics have programmed circuitry for data logging automatically and the data are periodically transported to the ground base via manual hook-ups. Therefore, the structure's condition can be monitored more frequently with even lower cost, compared with scheduled SMH, as shown in scenario B. The frequency of maintenance tasks in this scenario is assumed to increase by n times than that in scheduled SHM. The repair threshold should be increased to a certain extent accordingly. This additional procedure is called maintenance assessment. In order to maintain the same safety level as scheduled maintenance, scheduled SHM is requested at every scheduled maintenance cycle just as scheduled SMH and the threshold is adjusted to the previous damage size. The logic procedure of scheduled CBM is shown in Fig. 9.7.

Between every maintenance assessment, unscheduled repair is conducted as long as the damage size exceeds the threshold D1 so that the structure can be repaired without waiting for the next scheduled maintenance cycle. Otherwise, the structure with a large damage size will be kept in service, which may compromise the operational safety. It is noted that unscheduled repair is used here to distinguish from the repair performed at current scheduled intervals. However, since the threshold D1 in maintenance assessment is larger than D in scheduled maintenance, scheduled SHM should be carried out to repair structures with damages larger than D so that the structural safety is not impaired in the long term. This maintenance scenario can be seen as a hybrid model of scheduled SHM and CBM in order to maintain a high safety level.

D. CBM

The most advanced scenario is presented to achieve real-time monitoring, which is based on a mature on-board SHM system and a well-developed air-ground data link system. Data relevant to structural health is collected, transmitted, and processed continuously. It is important that the maintenance decision-making module can perform autonomously and inform the operator in a timely manner when to take maintenance measures. In other words, a complete CBM is realized. Since the structural health state can be monitored in real-time, the repair threshold can be also set to D1. However, considering the reliability of the SHM system itself, it is necessary to assess the system frequently. The CBM logic procedure is shown in Fig. 9.8.

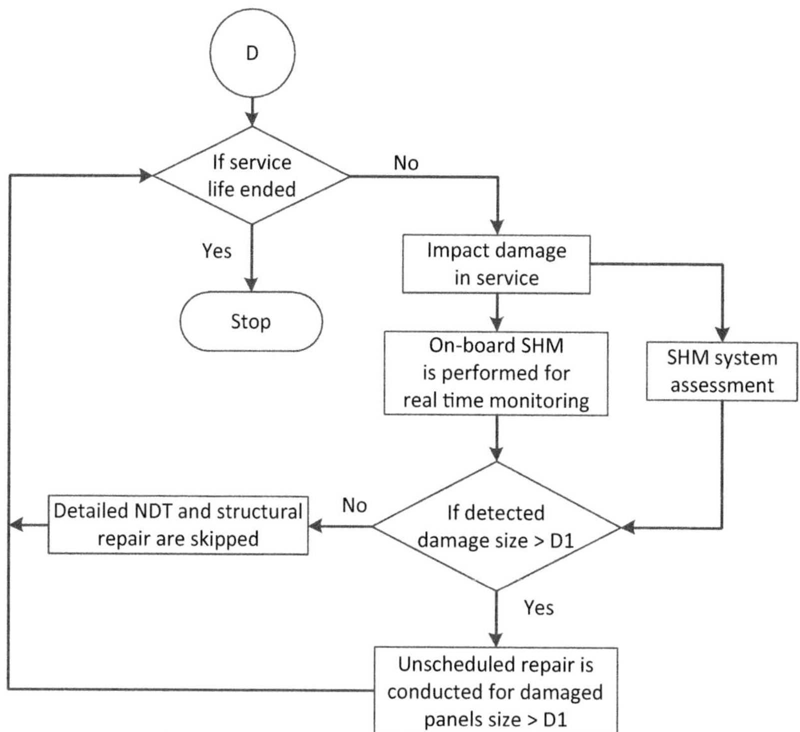

FIGURE 9.8 **CBM procedure.**

Overall, with SHM systems applied in the last three scenarios, structural damage can be detected by on-board sensors either at every predetermined interval or continuously, which spares operators great time and energy that would otherwise be spent on unnecessary intrusive inspections. Theoretically, new maintenance procedures incorporating SHM tasks have been developed. Quantitative analysis can be performed as long as detailed assumptions or realities are determined, such as the maintenance assessment or SHM assessment interval as well as the critical repair threshold. The maintenance cost and probability of failure of the composite panel might be obtained by using the probabilistic model developed in the previous chapters.

9.5 CONCLUSIONS

Aircraft manufacturers are promoting structural/system health monitoring to reduce long-term maintenance cost and increase aircraft availability. Efforts are underway to develop self-sufficient SHM systems for continuous monitoring, inspection, and damage detection in aircraft structures and systems in order to improve safety, reduce labour cost, and minimize human error.

Significant efforts, ranging from design to deployment, performance, and certification, are still needed to assure the safe incorporation of SHM into the highly regulated aviation industry. MSG-3 document is being revised by the Maintenance Programs Industry Group to incorporate the new technologies, such as SHM. This chapter proposed an example of how a combination of various SHM tasks can be integrated into the current MSG-3 logic analysis. More efforts are required to further the analysis logic for scheduled maintenance development. Future research directions include but are not limited to SHM system evaluation method to assess the SHM damage detection capabilities for ED/AD/FD, SHM system monitoring capabilities and inherent reliability of the system. When an automated SHM system is used to monitor structure that has maintenance/inspection requirements, appropriate transfer policies should be used to show that the intent of the damage detection (ED/AD/FD) requirement is being satisfied by the monitoring system.

Chapter 10

Maintenance Control and Management Optimization

Chapter Outline

10.1 INTRODUCTION

So far we have concentrated on engineering aspects of maintenance management, including the reliability of equipment and the origins of maintenance processes we have dealt with. Now we turn to the important human side of the maintenance function, the people on whom the processes depend: How do we prepare them for the work? What is their training and qualifications? How do we organize them and help look after them?

Reliability Based Aircraft Maintenance Optimization and Applications
http://dx.doi.org/10.1016/B978-0-12-812668-4.00010-1

Having gained a basic understanding of maintenance analysis and LSA processes from previous chapters, we now need to consider how the outcome is applied to operations and maintenance planning on the flight line and in hangars. Optimization of maintenance effort in support of the operations pattern is an important feature of this planning.

The objective of this session is to develop an understanding of the training, qualification, organization, occupational health, and safety management of the workforce of aircraft maintenance. The objective of this session is also to understand the processes of control of aircraft maintenance and development of a maintenance plan for an aircraft type, and how the outcomes contribute to effective management of fleet operations.

10.2 QUALIFICATIONS OF AIRCRAFT MAINTENANCE PERSONNEL

10.2.1 Educational Structure and Background

Aircraft trades fall within the recent government initiatives in establishing the National Framework for the Recognition of Training (NFROT), which establishes the recognition for all prior training. The Australian Qualifications Framework (AQF) provides a 13-level system of qualifications in postcompulsory education and training, from:

- Certificate level to Doctorate degrees
- The Higher Education Sector levels includes Diploma, Advanced Diploma, Bachelor
- Degree, Graduate Certificate, Graduate Diploma, Master's Degree, and Doctoral degree

The Vocational Education and Training Sector that we are largely concerned have the following six levels:

- Advanced Diploma—equivalent to 3 years full-time study post-VCE
- Diploma—equivalent to 2 years full-time study post-VCE
- Certificate IV
- Certificate III
- Certificate II
- Certificate I

The "trade" certificates indicate a graded level of competency upwards from I to IV, the latter being equivalent to a "technician" level in fields, such as electronics, NDT, or instrument calibration. In very general terms Level I is a basic entrant level, Level II is a qualified basic tradesperson and Level III is a skilled tradesperson. Specific tasks in the workplace will call for differing levels of skill: and part of the art and skill of management is to ensure task requirements and skill levels are properly matched. Control of this aspect of maintenance management is essential if the aims of aircraft airworthiness and workplace safety are to be assured.

10.2.2 International Requirements

The Licensed Aircraft Maintenance Engineer (LAME) has for decades been the international civil aviation trade standard. The requirements for licensing are based in the ICAO Chicago Convention, Annex 1, Chapter 4. This convention also provides for an aircraft operator to provide a maintenance manual, which includes inter alia "the responsibilities of the various classes of skilled maintenance personnel." The responsibility for standards required by licensed personnel for "certifying the aircraft as fit for release to service after maintenance" is vested in the state airworthiness authority, which is to ensure that "possession of an appropriate license demonstrates a level of knowledge and experience that may be appropriate as a basic qualification for certifying personnel."

10.2.3 Australian Civil Aviation Requirements

These are set out in various Civil Aviation Regulations. Section 30 provides a requirement for Certificates of Approval issued by CASA for inter alia "the maintenance of aircraft," "the maintenance of aircraft components," and "the maintenance of aircraft materials," together with a range of conditions that must be met, including the qualifications of personnel involved. CAR's Section 31 covers Aircraft Maintenance engineer licenses in one or more of the following five categories:

1. Airframes
2. Engines
3. Eadio
4. Electrical
5. Instruments

The license may, under Section 31, be endorsed for activities and types or categories of aircraft, and so forth. Licenses are obtainable on application, supported by documentation, which satisfies the licensing authority.

Under the National Aerospace Skills Project, in 1991 a new trade group alignment was agreed upon, based on up to seven levels of skill and three groups: mechanical, structures, and avionic.

Similar regulation applies to aircraft welders who need to be specifically approved for types of welding and parent metal groups.

10.2.4 RAAF Requirements

The RAAF maintains an aircraft trade structure very similar to the civil structure. However, there is a range of levels of qualification within each trade group with experience aligned to military rank. Pay levels are set on rank and trade group. There is a parallel but local system of recognition of training on an aircraft type. Certification for the release of aircraft for flight is not bestowed by a centrally based licensing system, but rather by local appointment to post in the organization having this delegated function.

A Technical Trades Review panel has recently revised the entire aircraft trades structure. There are now three broad trade groups similar to the new civil system, Aircraft, Avionics, and Structures, first introduced to the training system in 1992. Basic training is set at the Operating Level maintenance task, in line with the civil system for Aircraft Maintenance Engineer (AME). Additional courses are required to qualify for deeper maintenance employment. These changes have achieved significant training economies as well as facilitating civil recognition of qualifications.

As an example in the aircraft trades there are five levels. A person may be endorsed additionally as competent on Airframe and Engines:

- Aircraft mechanic
- Fitter
- Technician
- Advanced technician
- Systems technician

There are also separate trades in related skills of Aircraft Structures (was aircraft sheet/metal worker), NDI technician, surface finisher, aircraft and general welder, metal machinist, and aircraft life support (was safety equipment worker)

Most RAAF pre- and postreview aircraft technical trade qualifications have been recommended for accreditation and hence civil recognition through the ACT Vocational Training Authority.

10.3 SPECIFIC AIRCRAFT TYPE TRAINING

Most conversions to type training for maintenance tradespeople is conducted by the fleet operator. Courses may be tailored to specific segments of the workforce. With the relatively recent developments in training technology, many training courses can now be aided by synthetic training devices: for example, computer-based training (CBT) packages and simulation rigs. Much of this training can now be PC-based, although there can be value in networking workstations to allow for better training administration.

There are three levels of CBT:

1. Video, animation, demonstration, page turning
2. Programmed text testing, controlled response
3. System simulation, performance verification

These progressively deeper educationally interactive approaches give the student more feedback and realism in the interaction with the aircraft system and in problem solving. All allow for a degree of self-pacing by the student. Some are also more adaptable to distance education, which may be appropriate for some operational situations. Examples of such systems may be seen in the RAAF Training for the F-111 at Amberley and the Hornet F/A-18 CBT package at Williamstown; in both cases specific training rigs are set up to run courses on the various aircraft systems. In the Hornet case, an interactive software set can provide courses at level 2 (described earlier) with some system simulation rigs available for practical experience in various fault finding routines.

The more complex the training system is, in general the higher the development cost. As a broad indication, the level 2-type system is considered to cost about 400 h development time for each hour of training time.

UK experience argues that emulation of military aircraft systems using computer-based software rather than hardware simulators will provide the necessary mixture of formal theoretical training and hands-on exercises. A wide range of otherwise difficult faults can be effectively presented for solution. Benefits include relatively low cost, realism, maintainability, and interactive instructional monitoring.

Swedish experience of training on four generations of fighter aircraft systems conclude that, for modern integrated avionics systems with built-in test features, the only effective training is an interactive software-driven simulator and that a single all-systems maintainer properly trained may be the best solution, rather than a series of system specialists.

In summary it is evident that technology is being constantly developed to provide cost-effective training for the difficult challenge of maintaining modern integrated avionics systems driven aircraft.

10.4 OCCUPATIONAL HEALTH AND SAFETY

10.4.1 Introduction to Accident Control

The maintenance hangar or workshop presents a fertile field for the exercise of industrial health and safety management. The assets involved, both the aircraft equipment and the trained personnel, are of high value and the hazards presented for accident control, as safety practice is commonly called, are comprehensively available. We shall address the types of hazards, the potential outcomes of accidents that may occur, and then the management strategies that may be adopted in accident control.

The chain of events in most industrial accidents follows four steps (RAAF):

1. Failings of persons
2. Unsafe acts and conditions
3. Accidents
4. Damages, losses, injuries

Accidents in the workplace are of major concern to all involved, and in most cases are preventable. Accidents can have ramifications at five levels:

1. National—productivity and costs
2. Community—charges, shortages, taxation, and services effect
3. Worker—pain and suffering, pay, overtime, job change, confidence, social, and personal activities
4. Family—income, facilities, suffering, and disruption
5. Company/organization (see following paragraphs)

The serious consequences in the aviation field merit universal attention to management strategies for a safe workplace. The four broad steps required are:

1. Hazard identification in work area, work method, or in the worker
2. Control the danger—act
3. Prevent recurrence
4. Monitor

10.4.2 Hazard Identification

Hazards can be categorized into six types with a series of subcategories:

1. Physical: noise, vibration, lighting, electrical, hot and cold, dust, fire, guarding, and space.
2. Chemical
3. Ergonomic—design, handling
4. Radiation—X-ray, microwave, UV, IR, laser
5. Psychological
6. Biological—infection, bacterial, viral

Aviation provides rich opportunities for most of these. Some hazardous aspects are covered by formal Codes of Practice, typically:

- First aid in the workplace
- Manual handling
- Manual handling (occupational overuse syndrome)
- Noise
- Welding
- Synthetic mineral fibers
- Asbestos, and so forth.

Specific dangers fall into three main sectors:

- In the workplace:
  - Lack of order and cleanliness
  - Means of access

  - Storage and stacking
  - Moving objects
  - Surfaces and edges
  - Machines and equipment
  - Electrical installations
  - Harmful and flammable substances
  - Defective lighting and ventilation
  - Condition of roofs, walls, and structures
- In the work method:
  - Poor day-to-day maintenance
  - Unsuitable tools and materials
  - Poor and untidy layout
  - Handling materials
  - Unsuitable safety guards
  - Deficiencies in other protective equipment
- In the worker:
  - Knowledge of safety rules
  - Nonconformity with rules
  - Personal attire
  - General conduct
  - Divergence from approved methods
  - Manner of doing work
  - Use of protective clothing and devices

Hazard identification can be systematically processed, conducted by any party involved in the workplace or can result from some particular observation. In either case, good practice ensures that a hazard report in some form is completed on which follow-up action can be based.

10.5 ORGANIZATION FOR MAINTENANCE CONTROL

The functions of maintenance control are to implement the maintenance plan and coordinate and control all aircraft maintenance aspects to ensure that airworthiness of the operating fleet is achieved by conducting required maintenance tasks in an efficient and accountable manner.

In most organizations, control of maintenance flows along organizational lines, corresponding to the two maintenance levels, but with the addition of the "off-application" aspects of workshop maintenance, that is:

- Line maintenance
- Heavy (deeper) maintenance
- Workshop maintenance

10.6 SYSTEM OF CONTROL

The components of a reliable maintenance control system can be related to basic management principles as follows:

Planning: Details of all maintenance tasks required on an aircraft at any specific time must be accessible to the maintenance controller responsible for their management. Tasks including routine scheduled tasks, unscheduled tasks, repairs, modifications, and special inspections must be collated and coordinated into a work plan for each particular aircraft prior to work starting. Resources and time factors must be integrated into the plan so that the outcome meets engineering and operational objectives.

Organizing: Work teams must be briefed and adjustments to planned sequences made so workload is spread evenly and team commitment to the task sustained. Overtime requirements must be arranged as early as possible to retain harmonious personnel management.

Implementation: The servicing task should be conducted as close to the plan as is achievable. Unforeseen repairs or corrective maintenance actions should be able to be embodied in a revised plan in an efficient and effective manner. Each task has to be allocated to the tradesperson or group individually and clearly documented. The planned times of start and finish should be clearly set out. Completion of each task has to be recorded with the authorized tradesperson's certification together with details of the time taken for the task and its completion date and time.

Control: Records of completed servicing have to be compiled and reviewed to determine whether the plan was met, what variances occurred and whether changes are needed to ensure detailed planning for future servicing benefits from the experience gained. Records are required to be retained for a substantial time for reference, including as a historical traceability record for accident or incident investigation.

10.7 AIRCRAFT TAIL NUMBER MAINTENANCE PLANNING

The fulfillment of a long-term heavy maintenance plan in the most efficient way possible requires aircraft to have been flown to the maximum number of operating hours or cycles prior to being withdrawn from service for maintenance. There is an almost inevitable economic compromise needed between commercial or operational demands to meet schedules and plans and constraints of optimum long- and short-term maintenance plans. While airworthiness commitments cannot be compromised, some impact on efficient maintenance planning may be needed to avoid impacting operations and thus customer service. The manipulation of tail numbers to operational tasks is one mechanism for achieving optimum efficiency.

One parameter used by many airlines to measure effectiveness of their performance in this regard is the achieved mean operating time between scheduled maintenance as a percentage of the maximum allowable. The shortfall represents operating hours, which are lost to the organization when the aircraft is brought into maintenance before it needs to be. For short-term maintenance tasks, seeking to overnight the aircraft at an appropriate line maintenance station, rather than an airfield not so equipped, is one measure available to adjust operational planning in the interests of an economical maintenance plan.

Military maintenance engineers have traditionally relied on a display device called a "Stagger Chart" to give visibility to achieved flying hours against required hours to meet the progressive input of each aircraft into the long-term plan. Time slots for hangar maintenance are set up against the expected annual flying rate. A Gantt-type chart is then drawn up that compares the hours flown for each aircraft to the optimum hours to reach the time slot. Reallocation of particular aircraft tail numbers to operational tasks can then be made to adjust the achieved rate where possible to fly the aircraft into the required number of hours on the specified date for the maintenance slot. This chart will be further described during class.

Weekly operational flying programs are issued to take account of detailed aircraft operational requirements, which on average meet the authorized annual flying rate, which is controlled in peacetime by budgetary constraints. Operational inputs to the detailed tail number allocations may need to be negotiated to retain optimal planning against the overall stagger chart plan. Civil airlines are understood to operate a similar maintenance/operations planning process to achieve optimum economy in the maintenance plan.

10.8 CERTIFICATION OF WORK DONE

A fundamental tenet of aircraft maintenance control is the accountability of each individual for the decisions made and the work performed. In both civil and military systems plans and tasks must maintain integrity of the maintenance system from basic requirements to ultimate sign-off on the serviceable aircraft or component. The steps involved are as follows:

- Required tasks are identified. If a task is required as part of the approved maintenance plan then only an individual, properly authorized to do so, can vary the requirement, and they will be required to certify their decision in an accountable way.
- The task is placed in the maintenance section by an individual in the system authorized to do so and their action is recorded in an accountable way. The task may arise during flight or during another maintenance task and the requirement will be filled out by the aircrew or tradesperson involved on an accountable form.
- The maintenance manager of the section involved in doing the work also has to present an accountable worksheet or other instruction to the individual tradesperson who will eventually sign off as having satisfactorily completed the task.

- When all tasks have been so certified, the coordinator of servicing, having checked that all the work has been properly certified, will then sign off the servicing as completed and, if appropriate, certify the aircraft as airworthy.

10.9 MAINTENANCE FORMS

In most modern aircraft maintenance management systems, the documentation required for tasking and controlling maintenance work is produced by a computer-driven system. An understanding of the function of these forms illustrates the processes and accountabilities involved in maintenance control. The types of documentation tend to be relatively standard and comprise the following:

- *Aircraft maintenance form or maintenance release*: Used in line and hangar maintenance to record flying hours; changes of serviceability; certify maintenance actions; pilots' acceptance and release; corrective maintenance; replenishment, for example, refuelling, oxygen recharging, and so forth.
- *Maintenance notification or work release*: This form is filled by maintenance control to initiate a maintenance task not covered by a routine servicing. It is normally acquitted by an entry into the form but, if not, completion of the work is signed off on the form.
- *Servicing record certificate*: This is a tasking form to initiate and cover routine servicing scheduled requirements for aircraft and components and specifies the time date and details of the work to be done. It also records certification of completion of the work.
- *Unserviceabilities and component changes*: This is a method for the tradesperson to record the finding of unserviceability or for the maintenance control section to initiate a "hard-time" component change. Computer-generated tasks of this type will be generated by a similar form.
- *Aircraft log book, maintenance log, or tech log*: This provides an historical record of flying hours; maintenance actions; major or fatigue critical component changes, for example, engines, records of heavy landings, accidents weighings, and so forth., which could affect the structure, and any modifications incorporated. It may be maintained by hard-copy computer reports.
- *Aircraft engine log book or engine module log*: Similar to the aircraft log book showing operating hours, removals, test running, overhauls, modifications, and so forth.
- *Maintenance analysis and reporting system (MAARS) input report*: For items of maintenance significance, which are included in the maintenance plan for an equipment. This form reports the details of failures and maintenance action taken and supersedes the functions of the two following forms. It provides control of the serviceability of an item for the purposes of the computer-aided maintenance management system (CAMM) and the statistical inputs to the FMECA processes.

- *Serviceability tag*: A label, usually green, showing that the item it is attached to has been signed off as serviceable.
- *Unserviceability label*: Usually red, indicating a part not fit for use.

10.10 SERVICES

Replenishment and other services to support central maintenance functions are often closely linked to maintenance control. Some may be under a centralized control to spread resources effectively over a number of separate areas. Others may lie directly under control of the maintenance section manager. Some of these services are:

- Nondestructive inspection (NDI) support
- Spectrometric oil analysis (SOAP)
- Workshop support—machine shop, sheet metal shop, wheels and brakes, engine shop, hydraulics, avionics, and so forth
- Ground support equipment support and its maintenance workshop
- Motor transport, mobile cranes, fire tenders
- Refueling tankers and hydrant carts
- Gas cylinder support

Often there are special measures to provide visibility and control of the location and serviceability status of many of these important items of support. Sound communications and good responsiveness can make a substantial contribution to effective maintenance management.

10.11 MAINTENANCE SCHEDULES

The end-point of the maintenance analysis process is a table of maintenance tasks in a variety of forms, but which essentially set out:

- The maintenance plan, that is, process, required interval, and organizational responsibility
- Planned servicing schedules, that is, detailed tasks to be completed as part of scheduled servicing
- Maintenance and overhaul manuals, that is, covering processes other than planned services.

10.12 MAINTENANCE PLANNING

To explain the process, let us assume we have little first-hand experience on the first jet civil aircraft type for regular public transport. We have been provided with the manufacturer's Maintenance Requirements Document (MRD), approved during certification of the type, which establishes a set of periodic servicings and component maintenance tasks. These have been derived from the MSG-3 or LSA processes described previously.

In generalized discussions of maintenance planning, which may deal with many types of equipment other than aircraft, we would refer to our aircraft as "the application" and work done on it as "on application." Maintenance including engines and system components carried out away from the aircraft are classified as "off-application."

On this basis, airline management accepts our recommendation and decides to adopt the MRD in full for the first 12 months of operation. This is approved and published for use as our maintenance schedule. For convenience we will defer consideration of provisioning, purchasing, and related documentation consequences of reaching this stage until we deal with integrated logistic support in a section further on.

The structure of the approved maintenance schedule has two main parts: the scheduled services on application and those of the off application. Each servicing task has a reference code, related to the ATA nomenclature designating the system and item involved. This ATA 100 code is used widely throughout aircraft management for data control, including structuring of maintenance manuals and schedules. It is a three-element code. For example, 21-30-01 is illustrated as follows:

- 1st element-21-ATA Chapter/System—in this case Air-conditioning
- 2nd element-30-ATA Section/Sub-system—pressurisation control-may be up to 4 digits
- 3rd element-01-Sub-sub-systems breakdown—elements are assigned by the manufacturer.

A list of the full standardized code is as follows:
ATA System Classification

21 Air Conditioning
22 Auto Flight
23 Communications
24 Electrical Power
25 Equipment and Furnishings
26 Fire Protection
27 Flight Controls
28 Fuel
29 Hydraulic Power
30 Ice and Rain Protection
31 Instruments
32 Landing Gear
33 Lights
34 Navigation
35 Oxygen
36 Pneumatic
38 Water/Waste

49 APU
52 Doors
53 Fuselage
54 Nacelles/Pylons
55 Stabilizers
56 Windows
57 Wings
61 Propellers
71 Power Plant
72 Engine
73 Engine Fuel and Control
74 Ignition
75 Air
76 Engine Controls
77 Engine Indicating
78 Exhaust
79 Oil
80 Starting
81 Turbines
82 Water Injection

10.13 REFERENCE DATA DEFINITIONS

An aircraft system will have a series of scheduled servicings at specified hours. These may be based on flying hours, that is, the time from wheels-off to touchdown. Airlines and pilots for some purposes work to block hours (or chock-to-chock), which is the time from the start of motion under power to coming to rest that is, the time between terminal gates. The difference, the time for poststart checks and taxying time, is economically significant and it is important to be consistent in working to the applicable frame of reference for time.

Aircraft operating cycle: A complete take-off and landing sequence, touch and go landings are counted. Records of operating cycles are referred to for undercarriage and some structural limits.

Engine operating cycle: A completed engine thermal cycle, including application of take-off power. Consistent engine operating procedures governing power settings are assumed in these records and are worth monitoring for compliance.

Pressurization cycles: Records the application and release of pressurization loads on the fuselage to a specified level, which are significant for structural fatigue monitoring.

Some common abbreviations are time between overhauls (TBO), time in service (TIS), time since new (TSN), time since installation (TSI). These time records may be required to meet specified servicing intervals.

Useable-on-code: This type of code for an item indicates the model, variant, or next highest assembly of which the item is a part or to which it may be fitted. Some parts are superseded by configuration changes, for example, modifications, and this code is an important indicator of the correct fitment. The use of hierarchical codes in aircraft systems is not a simple process as the RAAF found in developing the LOAS structure, which developed from being related to a "list of applicable spares" coding to become a TMC "technical management code" in which a simple reference number, which has a singular applicability to a particular element in a system with strict rules as are frequently necessary to ensure integrity of data elements in critical database systems.

10.14 EXAMPLE OF AIRLINE MAINTENANCE SYSTEM DEVELOPMENT

10.14.1 Setting

In order to obtain a better understanding of the planning of a maintenance system, a slightly fictitious and simplified airline will be studied; Dada is from the Ansett Airlines system, which was in business for 60 years, but is now closed down.

Assume that the fleet consists of three aircraft:

1. 20 Fokker F27 aircraft
2. 12 Douglas DC9.30 aircraft;
3. 10 Boeing 727-100 aircraft.

10.14.2 Aircraft Checks

10.14.2.1 F27

This is a relatively old design and the system was not set up in the MRB fashion. Although heavy maintenance was to be carried out in Melbourne, very few units were based in Melbourne; some were based in Sydney, some in Brisbane, some in Adelaide, and one or two were based in Cairns. They usually returned to their bases at night and overnight maintenance was carried out there. Their interiors were less durable than those of later aircraft and as they were somewhat simpler types they did not carry as much duplication as later aircraft.

In the circumstances, it was judged necessary that these aircraft should return to the main base every 1500 h, which, at the utilizations achieved represented a return to main base every 7 to 8 months. Their block time utilizations averaged 2500 h/year, which represented a flying hour utilization of approximately 2250 h.

Note: Generally, all check periods mentioned are based on flying hours not block hours.

Each F27 has a full overhaul (check 8) every 18,000 h, a so-called half-overhaul (check 7) at 9,000 h, check 6s at 3,000 h and check 5s at 1,500 h. Thus, in an 8-year cycle, each F27 had 6 check 5s, 4 check 6s, 1 check 7 and 1 check 8; there were 20 aircraft and on average there were 15 check 5s per year, 10 check 6s, 20 check 7s, and 20 check 8s. Again, these are averages, as every aircraft was not used at the same rate. In all there were 30 checks of varying types each year.

During these checks components are removed as appropriate and replaced, in the main, by spare components usually previously overhauled. This applies to instruments, alternators, pumps, and so forth. as well as perhaps control surfaces or flaps, which can all be overhauled, but which will not hold up the return of the aircraft to service. Engines are treated as components and are only removed and overhauled, or checked as required, and are treated quite separately to the aircraft check system.

10.14.2.2 Boeing 727

The most economical way to carry out the checks is to use the MRB recommendations although, as mentioned previously these are unlikely to suit any but the largest airlines. You will remember the only checks that could put the aircraft out of service for a period were checks C, D, and E. Usually the check E could be incorporated as required with the check D, although corrosion checks, which are largely based on elapsed time, may in fact require a special check. It has been found that the check C can, usually very easily, be incorporated with overnight checks by carrying it out progressively. Thus it is important to concentrate work on the check D.

It is worth noting, in a major check of this type, that about 75% of the work involves disassembly and assembly of the aircraft, about 15% is usually defect rectification, and 10% is modifications. Thus breaking this check down usually means this disassembly and assembly has to be repeated, causing considerable economic penalty, although the evening out of the workload mitigates this to some extent.

The check D will, usually, after the initial operations, have a period of 8000 h. Block utilization is about 3900 h/year, or about 3750 flying hours. Due to the difficulties of planning and scheduling aircraft, the average achieved period between checks would be about 7500, that is, there would be a check D every 2 years, and for 10 aircraft there would be five check Ds per year.

10.14.2.3 DC9

The same comments apply here as applied to the Boeing 727, assuming again a check D at 8000 h, that is, 7500 h achieved and a utilization of 2500 flying hours per year means there would be a check D every 3 years; for 12 aircraft this would mean an average of 4 checks per year.

10.14.3 Work-Hours and Staffing Assessments

10.14.3.1 Work-Hours Per Check

In planning the maintenance system the only check work-hours required presently are those that are expended carrying out the check and then returning the aircraft to service. Those work-hours expended on overhauling or checking a component that has been replaced do not affect the system other than as sheer work volume, and it is then a matter of planning the complement of the component overhaul shop rather than setting off manpower against time out of service. The estimates of work-hours required under these circumstances are given here:

		Work-hours
F27	Check 5	800
	Check 6	800
	Check 7	3,200
	Check 8	4,800
B727	Check D	26,000
DC9	Check D	21,000

10.14.3.2 Work-Hours Available

Also required are the available work-hours per man per year. These people normally work a 5-day week with 4 weeks annual holiday and 10 public holidays per year. Thus the number of working days available are 365 – 104 – 20 – 10 = 231. On average, sickness accounts for another 6 days, leaving 225 days in all, and at 8 h/day that gives 1800 h per worker per year.

10.14.3.3 Shop Efficiency

The shop efficiency is also important; whereas on a short, well-planned task it may be possible to achieve an 85% efficiency; over a whole year 67% is more likely. This results from various unexpected problems, such as corroded bolts, a breakdown in spares supply, defective detail planning, delays in return of aircraft, and so forth.

10.14.3.4 Time Required

F27: Taking the F27 first, the average work-hours per year are:

$$15\times 800+10\times 800+2.5\times 3,200+2.5\times 4,800=40,000 \text{ work-hours.}$$

Allowing for the efficiency factor of 2/3 and for 75 h average annual overtime gives the number of staff required as:

$$\frac{40,000\times 3}{2\times 1,875}=32 \text{ staff}$$

Each check by itself could expect to have a greater efficiency factor—say 85% and overtime at the rate of say 10%. Thus a check 5 or 6 would take:

$$\frac{800}{0.85}\times\frac{1}{8(1+0.1)}\times\frac{1}{32}=3.34\text{ days}$$

That is, with an allowance for testing and inspection, 4 days.

The corresponding figures for the checks 7 and 8, which would possibly include a test flight, would be 14 and 21 working days respectively. All checks performed in a year add up to 188 working days.

Again with more staff used on the major checks, the out-of-service time drops. For instance, with 48 workers, the check 8 takes 14 working days. There is, of course, a limit here due to congestion of working space inside an aircraft as small as the F27.

Boeing 727 and DC9 It is necessary to carry out five check Ds on the 727 and four on the DC9 per year. If these must be contained within the 7.5 months quoted previously, then the staff required can be calculated as follows:

$$\text{Work-hours required} = 5\times 26{,}000 + 4\times 21{,}000 = 214{,}000\text{ work-hours}$$

Assuming staff working efficiency is 85% and have 10% overtime

$$\frac{214{,}000}{0.85\times 1.1\times 8\times 140}=204\text{ staff}$$

What is going to be done with these people for the remaining 4.5 months of the year? Some can be used on occasional doubling up of F27s or to accelerate F27 checks. There is no way enough F27s can be used at one time to occupy staff. During these checks some F27s must also be done at the same time; thus total workers employed would have to be considerably more than the figures mentioned before.

The above staff assessment does not takes into account the fragmented nature of the 7.5 months; without a doubt this would lead to considerable difficulty and increase the staffing required as well as leading to greater inflexibility as some larger checks could only be done in the larger periods available.

The system rapidly becomes both unworkable and uneconomic.

10.14.4 Further Development

This excessive workload must be reduced and there are three direct ways of doing so:

1. Reduction of the check D and spreading inspections elsewhere;
2. Extension of the check period;
3. Occasional doubling up of F27s

First, however, the 7.5 months previously mentioned should be examined. At that time, the separate periods for 1969 were 6 weeks, 1 month, 2 months, and 3 months. The last two are reasonably stable for other years, but the first two, while being the same in total, vary according to the date of Easter in the particular year.

Easter Sunday may occur on any date between March 23 and April 24. Thus the week of Easter travel will occupy the week between that of Wednesday, March 19 to March 26, and from April 20 to 27, with any variation between.

Assuming that February 10 and May 7 are fixed, which they will be within two or 3 days, the periods are either 36 and 41 days or 68 and 9 days, or any variation between. The sum in each case is 77 days, but over the years these can be split up in any way whatsoever and it is essential to have the greatest flexibility in the maintenance system to cope with this. In the latter case above, obviously that year is a short year for maintenance in that the 77 days is cut to 68 days, since the 9 days is of no real value for any form of major check on the Boeing 727 and DC9. The other periods of the year are for 67 and 90 days, respectively.

It is worth noting that if each check took 5 weeks, it would be possible to complete only 5 checks in any year; 4 weeks is not much better and is more wasteful of days, but 3 weeks enables 10 checks to be done each year, and is no more wasteful of days than for 5-week checks. Again, the shorter period gives more room to maneuver during the first two periods broken by Easter, whenever it falls.

Returning to the three methods of reducing or spreading the workload, reducing the size of the check will throw more checking onto the intermediate check. However, this must be one of the ways to proceed. Extension of the check period is only possible in light of experience and means some inspections cannot be extended and thus are thrown onto earlier checks. Doubling up on the number of F27s on check at the same time is possible and has to be adopted in some cases, although during heavy B727 checks F27 are kept out of the shop and all staff concentrates on the B727.

The final solution was to introduce a 5,000 h ± 1,000 check, and extend the major check period to 10,000 h. As far as possible, maximum flexibility is retained on the 5000-h check and the major work consists of modifications and rectifications. Essential checks required before 10,000 h are whenever possible fed into the overnight checks rather than the 5000-h check. Nevertheless, disassembly and assembly times dictate the total work-hours expended on the two checks is considerably more than would have been expended on the single 8000-h check, but the extension of the period alleviated this.

Brief mention has been made of the intermediate check being broken down and carried out overnight. This intermediate inspection is usually specified as every 1000 h. Instead, it was broken into four parts and joined up with the terminating flight inspection and overnight rectifications, and then carried out every 250 h; this itself could be carried over two or three nights. The terminology adopted by Ansett was checks 1, 2, 3, and 4, each of which contained one quarter of the 1000 hourly check, as well as occasionally having some of higher check inspections, when the check D is reduced to what Ansett called a check 8. The 5000 hourly inspection was called a check 7.

10.14.5 Direct Application

The solution adopted for the simplified Ansett Airlines example was to have a very much modified check D at 10,000 flying hours with an intermediate check at 5,000 flying hours, which was very much a rectification and modification block, rather than inspection with the remainder of the inspections being carried out overnight when necessary, having already spread the 1000-h check C into overnight checks. This meant of course expanding the workload at overnight ports, but provided this could be accomplished at various ports, there was minimum disruption to service.

The man-hours expended on these checks are given as follows (Table 10.1).

For the Boeing, assuming as before a flying hour utilization of 3750 and an achieved check period of just under 9400, two check 8s will be carried out in 5 years; there will also be in this period two check 7s. If there are 10 aircraft, this means 20 check 7s and 20 check 8s in 5 years, or four check 7s and four check 8s per year.

For the DC9, assuming as before a flying hour utilization of 2500, an achieved check period of just under 9400 would mean there would be four check 8s and four check 7s in 15 years, and thus for 12 aircraft there would be 3.2 check 7s and 3.2 check 8s per year. We can take this as three check 7s per year since the check 7 period is more elastic due to its low specified inspection content and thus the full period of 5000 flying hours can be achieved. In the case of the check 8, however, there would normally be three check 8s per year with an occasional extra one in some years.

The average workload per year is thus:

$$4 \times 20,000 + 4 \times 10,000 + 3.2 \times 16,000 + 3 \times 8,000 = 195,200 \text{ work-hours}$$

Plus, as before, 40,000 h for the F27, giving a total of 235,200 work-hours.

Again, assuming 1875 h per person per year, and an efficiency of 66.7%, we find that the number of workers required is 188.

TABLE 10.1 Check Hours

	DC9	B727
Check 7 (5,000)	8,000	10,000
Check 8 (10,000)	16,000	20,000

If we assume this is the number of people employed on this work, we can compile a table of workdays required for each of the various checks assuming:

1. There are no F27s in the shop at the same time;
2. There is one F27 being checked at the same time.

		Working days	
		No F27	One F27
Boeing 727	Check 8	15	18
	Check 7	8	9
Douglas DC9	Check 8	12	14
	Check 7	6	7

These days have been rounded up to the nearest day to allow some time for testing and in some cases flight test. By careful scheduling and always using the total workforce on the B727 check 8, it is possible to schedule this work within the 7.5 months mentioned before. If not, then extra staff would have to be employed. This conclusion can be drawn as follows:

In 7.5 months there are, on average, 162 working days. Whether this can be achieved can only be determined by a more detailed plan. To view this more closely, it is assumed the above fleet is to be maintained using the average occurrences as a basis for the plan.

Table 10.2 shows the number of working days required for each of the checks for both 188 workers and 156 workers. In addition, workdays required are shown for the F27 for both 32 and 48 workers.

First, a chart must be prepared for the whole year, showing periods of projected high traffic and also public holidays. In practice the chart will have a square for every day of the year and is quite bulky. Producing one of these would be unwieldy. Instead, a miniaturized one will be shown to illustrate the method.

The following rules are adopted:

1. All Boeing 727 check 8s occupy the full staff, that is, there are no F27s in at the same time, although there can be a small amount of overlap.
2. On any check, if the full staff is used, no F27s can be worked on.

TABLE 10.2 Summary of Check Hours

4 Check 8s	B727	60 working days
4 Check 7s	B727	36 working days
3 Check 8s	DC9	42 working days
3 Check 7s	DC9	21 working days
Total		159 working days

3. Check 8s for any aircraft of the same type must be separated by at least a month and preferably 6 weeks; this allows time for components replaced on a check 8 to be overhauled in time for the next check 8.
4. Whenever possible, aircraft come in on a Sunday night so work can proceed on Monday morning; this is particularly true for the F27 checks 5 and 6, so the aircraft can be returned to service on the following Friday morning.
5. When possible, aircraft should be returned to service on a Friday.
6. F27s can be doubled up if necessary during school holidays, preferably in the middle to avoid the high traffic at the beginning and end, and again for preference during the May holidays rather than in August through September.
7. The most popular holiday periods for staff are Christmas and January and the August through September School holidays.
8. Weekend work is avoided and only required in exceptional circumstances.

Again, if you remember, we required an average 2.5 checks 8 and 7 per year; for 1978, we assumed three check 8s and two check 7s.

This plan is not as practical as it might be—there is very little margin for error. Any delays affect the whole program and, in fact, a relatively small night shift would ease problems considerably. Total staff required would be unaltered, but in the main there would be two major effects:

1. The night shift would involve itself in sorting out bottlenecks.
2. The night shift would also be responsible for preparing an aircraft ready for full-scale work first thing the following morning. This refers to draining of fuel, jacking up, and preliminary opening up.
3. The result of (1) and (2) could well be an improvement in the short-term efficiency factor.
4. Again, during those periods on the F27 when there are no large aircraft in the shops, it would enable more staff to work on the aircraft during the 24 h, that is, work in congested areas could be staggered more easily.

You will notice no complications are assumed, as would be caused by the state of the fleet with reference to flying hours. These will be dealt with in more detail later, but either aircraft have to be pulled out early or a special extension has to be made, provided certain inspections are carried out. Normally, however, provided aircraft inspections are sequenced initially, it is possible to pull aircraft out of service reasonably near the prescribed period. There is usually an allowable tolerance; this, however, is why the achieved period is usually less than the prescribed period.

Chapter 11

The Methodologies of Reliability and Maintainability in the A380 Program

Chapter Outline

11.1 INTRODUCTION

In January 2005, the world's largest airliner completed its maiden flight on the runway at Toulouse Blagnac airport in the South of France [96]. As the flagship of the 21st century, the 555-seat A380 is not only the biggest civil aircraft ever built, but also the most advanced, demonstrating a unique technology platform from which future commercial aircraft programs will evolve. It sets new standards in air traffic and, at the same time, opens a new age in commercial aircraft reliability and maintainability (R&M) design.

The key reliability target for the A380 is much higher target Airbus had on its A340 series, which was 99% in-service operational reliability (OR) within two years of entry-into-service (EIS) [97]. It is obvious that when very large aircraft (VLA), such as the A380 get larger and the range of flight gets longer, it is more challenging to meet the marketing requirements for operating economics. Therefore, R&M has been placed high on the list of A380 design priorities [98].

Over the past two decades, aircraft technical operation capabilities have received extensive attention, especially with growing interest in flight safety, reliability, maintainability, and so forth. In terms of economical efficiency of operations, aircraft safety and reliability must be raised to a higher level at all development stages of the aircraft life cycle. This is to be done through more sophisticated R&M optimization methodologies.

Reliability Based Aircraft Maintenance Optimization and Applications
http://dx.doi.org/10.1016/B978-0-12-812668-4.00011-3

It is the first time that an aircraft level reliability/safety process has been implemented on a large-scale commercial aircraft (i.e., Airbus A380). The primary objective is to identify aircraft level functions and associated failure conditions; decompose them to multiple systems, and place requirements on various levels of suppliers in the development of systems and equipment. The aircraft level reliability analysis provides the basis for the system level reliability analysis and is maintained throughout the entire lifecycle of the aircraft. They serve to drive the design to improve system/component reliability toward the achievement of the defined target values.

Conventional methods in the field of reliability analysis have been shown to be insufficient for the increasing complexity of aviation design requirements. For example, current reliability prediction techniques are not really representative of real situations. Very often, through those aircraft reliability assessment/prediction methods, it can hardly indicate how the reliability is achieved in the actual scenarios. There is a substantial need to transform from conventional prescriptive reliability prediction techniques to a more generic approach of reliability modeling methodologies, which can be able to conclude all aspects of the product life cycle [99]. For instance, when calculating the OR of the aircraft, according to the MBR documents, all the related maintenance checks and their intervals should be taken into consideration as one important factor contributing to the overall reliability of the aircraft throughout its service life. The Markov techniques are in a better position to handle redundancy management, operational considerations, and safety issues than the conventional methods. At this stage, the Markov chain method has been employed to carry out the reliability modeling for A380.

For the first time in Airbus's history, maintenance engineers have been integrated into the aircraft design team from the very beginning of the A380 program [26]. This approach has been implemented to develop a statistical model for further reliability enhancement. The Bayesian theory has been utilized to combine expert judgment on engineering concerns with service data on previous similar aircrafts. The primary goal of this method is to address the most important engineering concerns to be relieved.

The overall reliability design of the A380 was assessed and analyzed through the validation and verification (V&V) process [100]. It has been shown that the outcome of the V&V process is the important basis on which all critical engineering decisions need to be made. A conceptual study of this process is presented in this chapter together with the case study of the A380 program.

In terms of operational economics, the maintainability of the aircraft is all about controlling maintenance costs inducing from the inherent aircraft characteristics (i.e., aircraft reliability). Lots of innovations have been brought onboard the A380; one key objective is to reduce the direct maintenance cost (DMC). As a matter of fact, those innovations are the products of the comprehensive consideration of A380's maintainability design and optimization, which will be further discussed.

11.2 RELIABILITY MODELING APPROACH

Aircraft development process at Airbus is currently based on concurrent engineering (CE) or simultaneous engineering (SE) principle, this principle is defined as a systematic approach to the integrated concurrent design of products and their related processes, including manufacturing and product support [101]. CE process was also successfully implemented in the Boeing 777 program delivered in mid-1990 [102]. CE aims at interweaving consecutive development steps by breaking down the overall product development into independent work packages, which are all managed in parallel. Those work packages cover aircraft performance and operational characteristics. This is done in order to reduce the aircraft development cycle as much as possible. One of the key significances out of the CE principle is that the in-service performance of the aircraft including inherent reliability and OR has to be predicted even earlier in the product design stage so that the direction of the product development can be truly driven by the customer requirements, thus product revisal and failures can be minimized. Normally, there are three typical reliability predicting methods widely used in this field:

1. Classic FTA and RBD combined methods;
2. Monte Carlo simulation method;
3. Markov chain modeling method.

For the first method, it is a relatively simple way to implement the reliability modeling, but it is hardly to take various types of dependencies into consideration at the same time, such as repairs, (scheduled) maintenance, coincident failures, redundant/spare systems, and so forth. For complex systems, such as aircraft, using this method will result in very complicated reliability models (i.e., FTA and RBD). Meanwhile, one significant disadvantage of the Monte-Carlo simulation is that an enormous amount of simulation cycles are required for complex systems due to the demands for more accurate results. Additionally, many aerospace companies continue to use MIL Handbook 217 for reliability predictions, which are highly dependent on in-service data of the aircraft. However, there is often not enough in-service data at the correct level to enable a qualified safety analysis. On the other hand, the Markov chain method can cover a wide range of system characteristics because it doesn't have those limitations. A Markov model can even comprise factors including airline maintenance policies, dispatch requirements, air safety considerations, airworthiness standards, and so forth [103].

The advantages of the Markov chain method have been used to establish a generic reliability model to predict the A380 OR provided by each item of equipment and component with its own parameter values given. These values include MTBF, MTTF, MTTR, MEL application rates, mean DMC per flight hour, and so forth.

The Markov chain method or a Markov process, named after Russian mathematician Andrey Markov, is a stochastic process, which means that given the present state, future states are independent of the past states. In other words, the description of the present state fully captures all the information that could affect the future evolution of the process. Future states will be reached through a process, that is, a probabilistic instead of a deterministic one. During this process, the system may change its state from the current state to another state, or remain in the same state, according to a certain probability distribution. The changes of the states are called "transitions," and the probabilities associated with various state-changes are called transition probabilities. The Markov model is a graphical probabilistic model that describes transitions between states. Normally, the mathematical expression of the Markov chain theory is:

$$\Pr(X_{n+1} = x \mid X_n = x_n, \ldots, X_1 = x_1) = \Pr(X_{n+1} = x \mid X_n = x_n) \qquad (11.1)$$

where X_1, X_2, X_3, . . . are a sequence of random variables (e.g., parameter values of the aircraft) with Markov property. It has been recognized that Markov processes can be used to adequately estimate the OR of complicated aerospace systems [104].

The definition of the term aircraft OR can be described as the number of times per 100 takeoffs that the aircraft fails to take off within 15 min after the scheduled departure time. OR is then a probabilistic value to be predicted during the design stage of the aircraft. The prediction of the OR is based on the estimation of the probability that the aircraft may not be in a proper state to take off 15 min after the scheduled departure time due to aircraft technological problems (excluding administration and other events not related to technical failure). The prediction is to be broken down to a system level. The reliability at the aircraft level is then estimated by the sum of the elementary probabilities on all of its systems (e.g., mechanical, hydraulic, avionics, electrical systems).

In the early design stage of the Airbus A380 program, top level aircraft requirements (TLAR) for OR have been established and apportioned down to the level of systems, subsystems, and components [98]. This is an iterative process that travels along the entire process of the aircraft design, which shares some characteristics with some typical reliability analysis processes (e.g., FMECA or FTA).

An example is given below to illustrate the way to define the transitions and states of a system. Assuming a particular system, subsystem, or component (electrical or mechanical), all the possible states (expressed as GRP) of the aircraft affected by the condition of this equipment can be identified. The mathematical relations then can be expressed as:

$$\begin{aligned} \mathrm{GRP} &= \{\mathrm{GRP}_1 + \mathrm{GRP}_2\} \\ \mathrm{GRP}_1 &= \{\mathrm{STA}_1 + \mathrm{STA}_2 + \mathrm{STA}_3 + \mathrm{STA}_4\} \end{aligned} \qquad (11.2)$$

It is assumed that only four of those states can enable the aircraft to take off within the scheduled departure time. The four states ($STA_{1\text{–}4}$) have constituted the group one expressed as GRP_1 in the Eq. (11.2). The states of not being able to take off during the mission due to the malfunction of this equipment have constituted the group two as expressed in GRP_2. Those four states are defined as:

STA_1:	Full OK	This equipment is fully functional without any problem.
STA_2:	Full OK+	A problem has been detected but not correctly solved.
STA_3:	MEL GO	A residual problem has been detected and the MEL has been correctly applied to take off.
STA_4:	MEL GO+	A residual problem has not been solved and the MEL has not been correctly applied to take off.

The states of the equipment are described by more than two values (OK/not OK) containing single, double or multiple complicated failures. Fig. 11.1 illustrates the generic Markov process from the state STA_1 (Full OK). It describes the possible evolution of the aircraft's state from the first flight mission, through the ground phase, and to the beginning of the next flight mission.

Referring to Fig. 11.1, after the first flight, if there is no fault found, then the aircraft is ready for the next flight mission, which is recorded as STA_1. If a problem is detected during the first flight/ground time, two subsequent actions are involved: fixing the problem and checking the MEL. There will be four different results after those actions, which in this case are the four states defined previously. If the problem is correctly solved, then the aircraft is able to move on to the next flight (STA_1). If a problem is found and is not correctly solved, the aircraft may still continue the flight mission as the equipment with the problem is not in the MEL (STA_2). If a residual problem detected is not in the MEL, the aircraft may go straight to the next flight without any correction of this problem (STA_3). And if the MEL has been incorrectly applied, although the detected

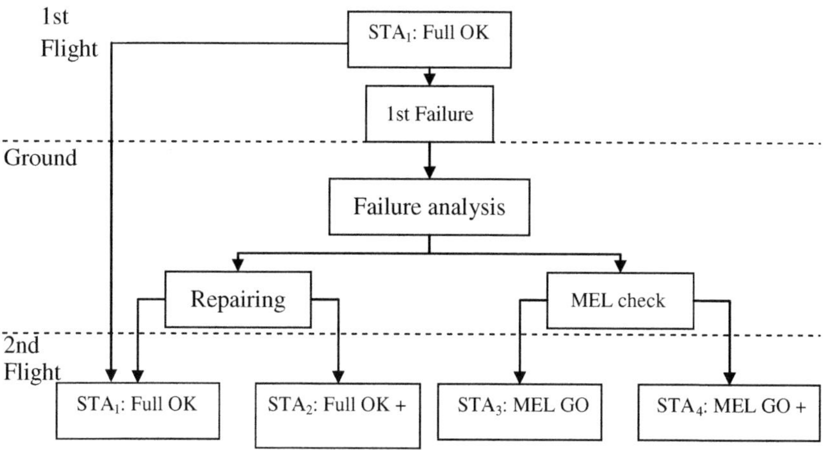

FIGURE 11.1 **Evolution from state "FULL OK" [105].**

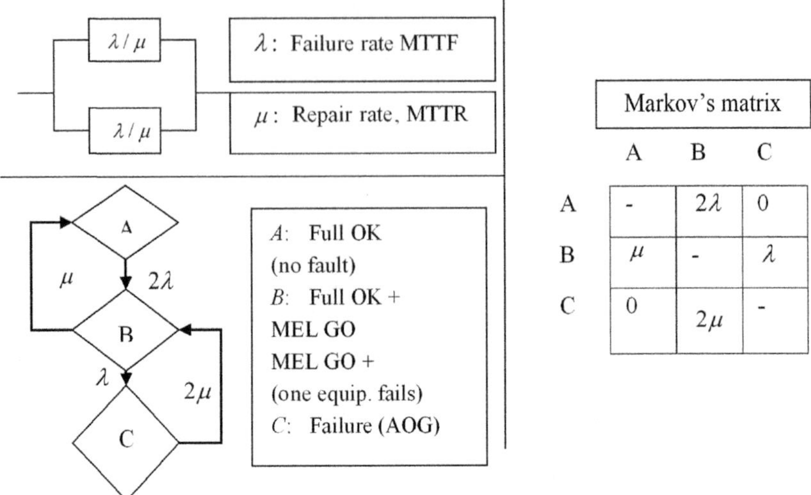

FIGURE 11.2 Illustration of Markov processing and modeling method.

problem has not been solved, the aircraft may still take off. In any other cases (GRP_2), the aircraft then cannot take off but instead is grounded.

The transition rates between all the states are defined from the mean remaining time in a state and a discrete probability transition law, which reflects the assumption that the transitions of the states between two distinct times—t_1 and t_2—should depend only on elapsed time (t_2–t_1) and on the probability state redistribution law applicable at time t_1. This assumption is based on a widely accepted hypothesis for complex system reliability analysis [106]. Furthermore, exponential distributions are assumed for all transition laws.

Specific software can be used to integrate all the Markov processes to determine the overall OR. A sample of the Markov processing and modeling method is illustrated in Fig. 11.2. Here the aircraft is assumed to be a simple redundant system with two parallel sets of equipment. The overall reliability is then calculated by solving the Markov's matrix.

Matrix manipulations are very tedious and complicated. It is hardly possible to solve by hand even for a simple system. However, numerous Markov chain solvers are available. With the fast-developing information technology, computer-based Markov tolls have been successfully applied to many large and complex systems by aerospace companies, such as Airbus, Boeing, and NASA.

The aim of the process using the Markov chain method is first to calculate the probability that the aircraft cannot depart due to a particular system, and then integrate all the subprocesses as a whole to predict the aircraft level OR on the aircraft level. The Markov process model enables analysis of a large number of complicated multistate transitions, including factors related to human aspects.

In the Airbus A380 project, two reliability design methods are carried out simultaneously to ensure the aircraft level target can be achieved [98]. For both methods, if targets are not met then the causes and objects are identified with corresponding design actions required to improve the reliability or to reduce the rectification time.

The first method (as discussed in this section) uses extensive availability models based on predicting system malfunctions, which lead to potential delays or cancellations, and rectification times. The objective of the method is to ensure that the sum of the probabilities of those occurrences for which rectification exceeds the planned downtime meets the desirable requirements. The principle of the reliability approach modeling implemented in this method is based on the approach of the Markov process, as discussed previously.

The second method is based on the system safety/reliability assessment approach where probabilities are assigned to potential failure conditions likely to result in flight or ground interruptions. This method, which will be discussed in the next section, is a proactive reliability approach combining expert judgment on engineering concerns with service data on previous similar designs to further enhance the reliability of the aircraft.

In addition, designing A380 for OR also followed a validation and verification (V&V) process, in which the OR targets and requirements were defined at aircraft level and then broken down to ATA (Aircraft Transportation Association) specification chapters and equipment/component level validation. Then from equipment level back to the aircraft level, the verification was performed using simulation tools and accelerated reliability test beds, throughout the aircraft development and manufacturing process. This will be further discussed in later section.

11.3 RELIABILITY ENHANCEMENT PROCESS

The stringent performance standards of the A380 together with the increasing demands for higher availability at minimum cost, have confirmed the importance of consideration of reliability and maintainability (R&M) throughout the design of A380.

When considering R&M in parallel, component reliability must be coherent with the maintainability and maintenance cost objectives. One Airbus policy upon this issue is to place highly demanding targets on A380 equipment suppliers, particularly for components that have underperformed in the past [98]. Airbus has implemented a proactive holistic approach (A380 system safety/reliability assessment process) to the design of this super jumbo aircraft. This reliability enhancement process provides engineers with reliability information and lessons learned from previous aircraft maintenance experiences. The A380 program is the first Airbus project that integrates maintenance engineers into the aircraft engineering team, and in fact has done so from the very beginning [26].

When designing the A380, the design team has taken the pedigree of previous/existing designs into account. The concentration at this stage is therefore

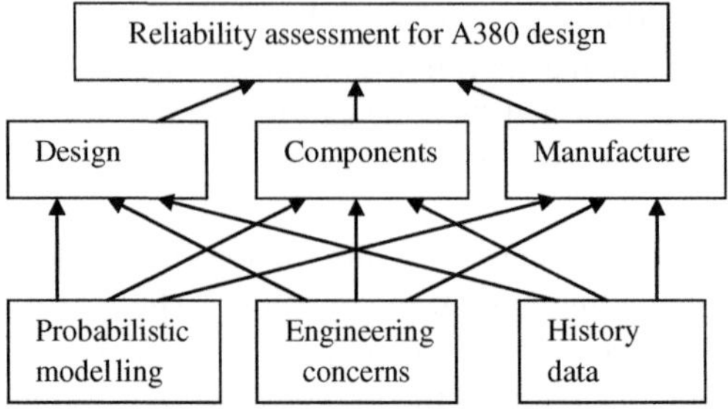

FIGURE 11.3 Reliability enhancement process formulation.

on the differences between the A380 and any similar designs. This process has utilized the Bayesian theory to combine the mathematical prediction, engineering considerations and service data on existing similar products together and analyze them as a whole [99]. Reliability models have been constructed for all (major) systems (e.g., engine, FCS) based on their history, current performance margins, and unscheduled failure distributions.

The outcomes from the process include a reliability function for the new design and a list of engineering concerns. The in-service/history data and engineering concerns are categorized and therefore the reliability function is obtained for all classifications (including design, components, and manufacture) and eventually for the whole aircraft. The results are used to compare the predictions/estimations with the initial design scope and the customer requirements. The purpose is to aid reliability enhancement by addressing the most important concerns to be mitigated [107]. Fig. 11.3 shows a simple illustration of this process.

In this paper, one Goodrich project is discussed below as an example [108], concerning the design of the primary Flight Control System (FCS) motor drive electronics for A380.

Some of the components in the A380's FCS are very similar to previous design configurations, but the power electronics and gate drive board were completely revolutionary designs for Goodrich. To apply the system safety/reliability assessment process, the initial decision made was the selection of historical data that most closely related to this new system. Since the A380 is the first commercial aircraft to use 5000 psi hydraulic systems for actuation of the FCS, there was no direct precedent in terms of functionality and operation environment. Goodrich finally decided to use the data obtained from Trent 800 and 700, because some control, monitoring, and power supply aspects of the Trent 900 actually evolved from those two earlier engine models. Engineering concerns were collected during the second stage. This stage consists of two steps, which are the conceptual design phase and the detailed design phase. The engineering concerns were raised in relation to the following systems:

- Power electrics
- Controller board

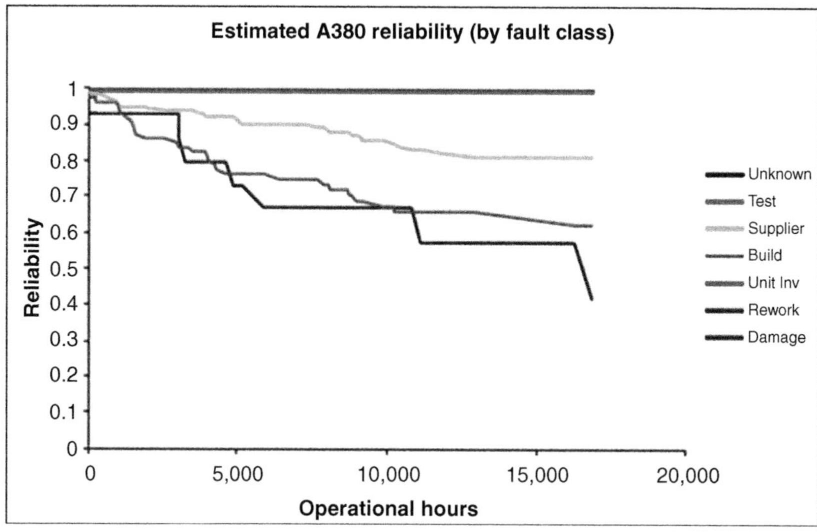

FIGURE 11.4 **Airbus A380 reliability estimation [108].**

- Monitor board
- Gate drive board
- Power supply
- EMC filtering

A set of engineers with the required full range of expertise was chosen from the project team. During the conceptual design phase, the engineers were interviewed individually. Some specific questions were addressed. For instance, the questions included concerns about any aspects of the system, the possibility that the concerns would cause a failure, and what mitigation actions could be taken to prevent those failures. After the individual concerns were gathered and analyzed, outstanding aspects were discussed by the group to identify the most critical concerns. Those concerns were then classified in the same way that the service/history data had been classified, to ensure the unification of the entire process. The final report was issued to the project team for future review and further study. As one important outcome, a statistical model was developed as the safety/reliability assessment for this particular system. An example of this is shown in Fig. 11.4.

11.4 VALIDATION AND VERIFICATION PROCESS

As discussed previously, the reliability of the A380 is modeled and assessed through the processes based on complicated mathematical theories. Critical engineering decisions are made on the basis of the predictions from those processes. A critical problem then naturally arises: how reliable are those predictions? The validation and verification (V&V) process has been introduced to address this question. Herein, the "validation" can be defined as a process determining whether the mathematical model describes sufficiently well the reality with respect to the decision to be made [109]. As a matter of fact, the mathematical models only transform the available information into the prediction; the information is limited although Airbus has made a huge effort to collect as much useful information as they can through sophisticated approaches of reliability modeling and assessment methodologies. In other words, the reliability of the predictions only depends on the quality of the input information. The mathematical problem is then solved by a numerical approach, which creates a computational model. This leads to the concept of "verification," which is a process of determining whether the computational model and its implementation result in prediction with sufficient accuracy. Fig. 11.5 shows the basic approach of the V&V process.

Normally, the validation pyramid of experiments is applied through the validation process. For example, Fig. 11.6 shows an idealized validation pyramid,

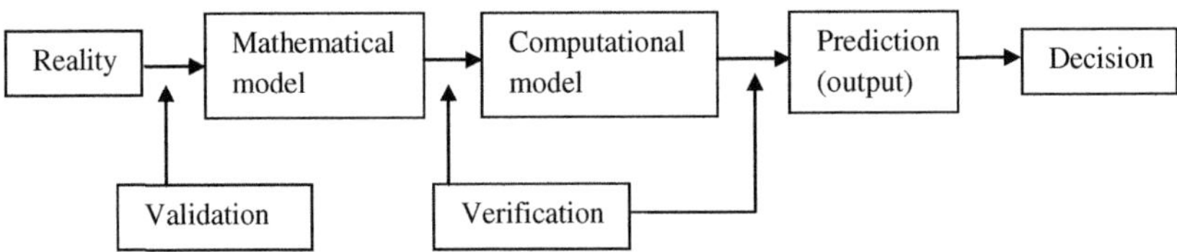

FIGURE 11.5 Scheme of V&V process in computational science [109].

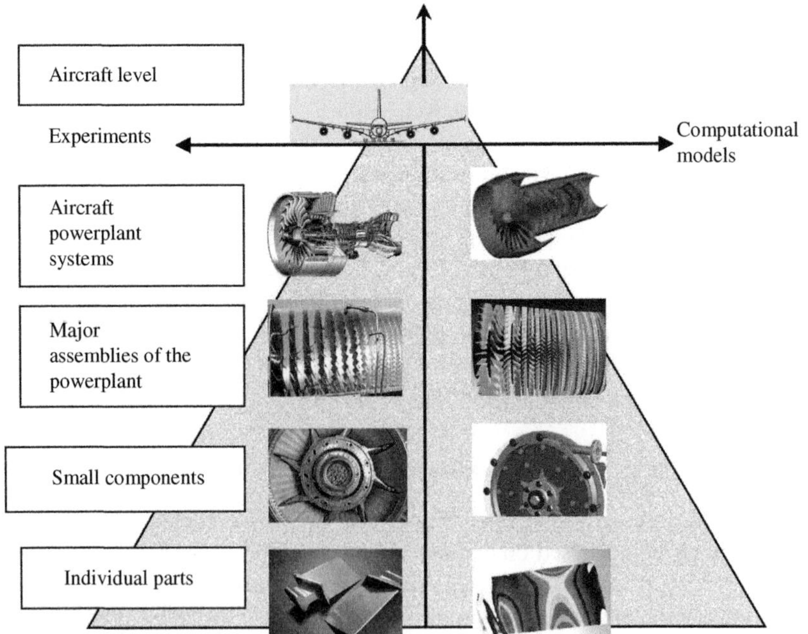

FIGURE 11.6 Idealized validation pyramid related to an aircraft engine design.

Flight-test aircraft	
MSN 001	First test aircraft, to be retained by Airbus
MSN 002	Right test aircraft. Third aircraft to fly. Will be delivered to Hamburg later this year and will be used first for evacuation tests, then fitted with working cabin (including Thales IFE) for "early long flights" and then for route proving. Will be refurbished and delivered to Etihad.
MSN 003	First weight-conforming aircraft and first aircraft for Singapore Airlines
MSN 004	Flight-test aircraft. Second A380 to fly. Workhorse for performance trials. Will be refurbished for Etihad.
MSN 005	Customer aircraft
MSN 006	Customer aircraft
MSN 007	Fourth test aircraft to fly, second with working cabin (Panasonic IFE). Will be used for route proving, and later re-engined with GP7200 engines to join MSN 009 for GP7200 route-proving. Will revert to Trent power and be refurbished for Etihad.
MSN 008	Singapore Airlines
MSN 009	Right-test aircraft. Leads certification with GP7200 engines. Will be re-engined and refurbished for Etihad.

FIGURE 11.7 **A380 flight test missions [111].**

which is related to the A380 power plant systems design, consisting of experiments and computational models, respectively.

As shown in Fig. 11.6, on the left-hand side are the experiments and on the right-hand side are the computational models. At the lowest level of the pyramid are the simple calibration experiments; on the highest level are complicated experiments and their computational analysis. Some of them are called accreditation (certification) experiments and serve as the basis for the demonstration of compliance with regulatory requirements (Fig. 11.7).

The comparison between the experimental (validation) data and the computed data is based on a specific metric (i.e., how the difference is measured) and the rejection criterion, which is a quantitative measure of the difference. The metric and the criterion have to be directly related to the prediction and the decision based on it. If the criterion is larger than the given tolerance, which is related to some threshold conditions, the model will be rejected. If the model at any level of the pyramid is rejected, then the model has to be changed. The revised model would need to pass all the lower level tests, and it is possible that more experiments would be needed. If the model is not rejected at a certain level of the validation pyramid, then the higher level is performed (Fig. 11.8).

The used tolerance is not arbitrary. It relates to the required accuracy of the prediction. If the required accuracy is low, then the tolerance could be large so that even a very crude model will not be rejected. If the desired accuracy is high, then the tolerance has to be small and many models could be rejected. The tolerance has to be chosen reasonably; otherwise, any practical model could possibly be rejected. If more than one model are calibrated and validated, then the best model could be possibly chosen and the tolerance adjusted so that the model will not be rejected. This can be done only if the adjustment is admissible

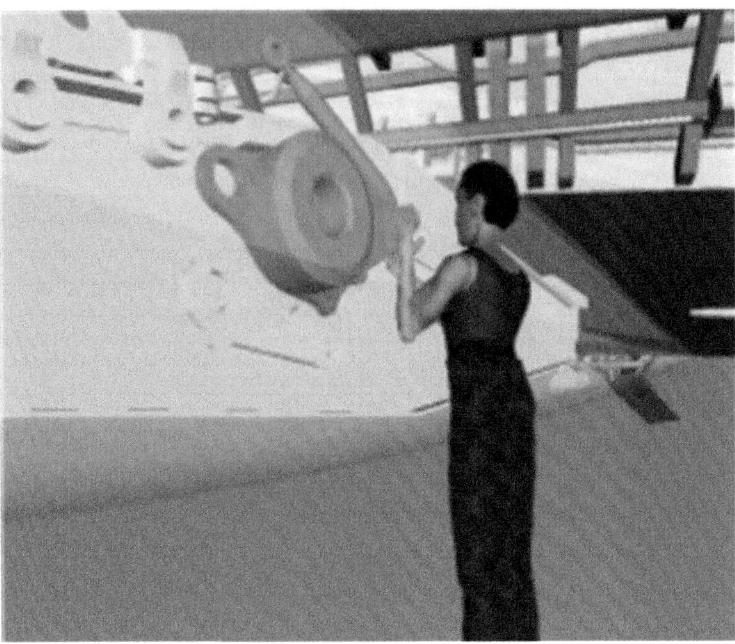

FIGURE 11.8 **Digital mock-up illustration.**

for the decision, based on the prediction. If the model is rejected, then a new model has to be created.

For instance, during the Airbus A380 wing loading test, the accreditation wing test failed on February 14, 2006 [109]. European Aviation Safety Agency (EASA) specifies that a wing in the static test must endure a load, that is, 150% of the limit load (worst-case scenario metric) for 3 s. The accreditation wing broke at a point between the inboard and outboard engines at 147% of the limit load. Some adjustment of the wing design was expected. However, this was within 3% of the 150% target, which has shown the high accuracy of the finite element analysis (FEA).

The A380 simulations and tests are two key contributions to adequate validation and verification. According to the aircraft, system, and equipment specifications, virtual simulations techniques are used to verify the integration of the A380 system operations. Various simulators are involved in the process, including desktop simulators, system integration bench, full aircraft simulator, and the full flight simulator.

With previous programs, such as with the A340-500/-600, there were no demonstration (tests) at aircraft level. But today, for the A380, Airbus has just one standard model, and it also has a demonstration at the aircraft level, which is an important contribution to the iterative V&V process mentioned above. There is a greater amount of testing than was ever done prior to EIS. This is also achieved through numerous full-scale test rigs and a series comprehensive flight tests (Fig. 11.7). These test rigs include "cabin zero" with an associated internal field emission (IFE) test-rig in Hamburg; "landing gear zero" in Filton, United Kingdom; the "iron bird" (for the actuation and hydraulic systems) in Toulouse; a fuel test-rig; and so forth [110].

11.5 ADVANCED MAINTAINABILITY OPTIMIZATION

Aircraft maintainability is all about controlling maintenance expenditures arising from the inherent design characteristics (i.e., reliability). Lots of innovations have been brought onboard to this cutting-edge super jumbo aircraft; one key objective is to reduce the DMC. Therefore, those innovations are the products of the comprehensive consideration of A380's maintainability design.

To support service reliability of the A380, the operators/airlines demanded much better monitoring of the aircraft systems in every aspect. To date, Airbus has offered systems monitoring as an option, but on the A380 it's a "set-in-stone" requirement. As a result, the aircraft features a lot of additional health monitoring sensors and software to observe every system in much greater detail. Initial demands focused on monitoring the engines (e.g., the QUICK technology for RR Trent 900 engines) and auxiliary power unit (APU) [97], but demands have since expanded to include other systems, such as the cabin air-conditioning packs. In this case, sensors have been installed to monitor the compressor outlet temperature. Contamination build-up on the heat exchangers affects this temperature and thereby signals the need for preventative action.

Still further development would be required to optimize the monitoring system, and this would likely continue beyond service-entry to further improve its effectiveness. For example, some Structural Health Monitoring (SHM) technology-based sensors are initially installed in some "hot spots" in the aircraft structure to obtain a practical database to validate and verify the role of this emerging structural monitoring technology in the future aircraft structural maintenance.

Meanwhile, Airbus has approached its component suppliers and asked them to set timed intervals for the maintenance tasks their products would require during their in-service life, with a proviso that they could not be below an Airbus-specified limit. As a result, Airbus will provide a list of maintenance tasks with intervals in "usage parameters" from which airlines can devise their own checks and inspections. Parameters have been set in flight hours, flight cycles, or calendar days [112]. All check-intervals have been defined referring to the MRB documents.

When A380 comes into actual airline operation, as an unavoidable fact, failures will still occur throughout the operational lifetime. Therefore, it is vital and necessary to build systems that will quickly and unambiguously identify a problem when one does occur. Moreover, once a failure has been identified, it is important that maintenance action can be conducted quickly without delaying a flight. Besides better systems monitoring, the A380 offers three further advances in troubleshooting technology. These include improved built-in test equipment (BITE), automated fault reporting through ACARS satellite datalinks, and free online access to Airbus's own troubleshooting software, Airman 2000 [26].

Until now, a typical BITE system offered only fault classification and display, but the A380 has interactive BITE, which makes current systems seem almost antiquated. The information can be presented on the ground and also to the flight crew through OMS. On a current aircraft type, for example, a system fault-notification usually will pinpoint a specific component. The troubleshooting manual will recommend a test to confirm the failure and refer the engineer to the aircraft maintenance manual for instructions on how to carry out the test. This might involve pulling a circuit breaker or turning on the hydraulic pressure and so forth, before the test can be run to confirm the failure.

With interactive BITE, the engineer simply presses a button to run a test and the aircraft configures the system for the requested test. This means that everything needed is already on the aircraft and no paper documentation is required. All required manuals are stored and interlinked in the central maintenance system (CMS) while online links to Airman 2000 are available through ACARS. Airman 2000 will be used extensively on A380 flight-test aircraft to collect statistical data. Airbus will provide this data to all A380 operators at service entry.

As airlines requested, maintainers' health and safety was considered one of the several priorities for the maintainability of the A380. For example, any equipment weighing more than 25 kilograms either requires two people to handle it or a specified device to carry it. All such components have hoist attachment provisions. Airbus also tried to optimize ergonomics around and inside the aircraft.

Those targets and requirements have been set accordingly and more extensive use of computer-aided design in the form of space allocation models and digital mock-up (DMU) simulating life-size mechanics are being effectively used to assist the design in coming up with the right maintainability solutions at an early stage. Hard mock-ups are used for specific issues to validate the simulation scenarios. The maintainability qualities, together with the tooling and maintenance practices will be subject to verification and airline demonstration before entry into service.

The OEM devised a sophisticated virtual mannequin to conduct maintenance tasks in its aircraft DMU, which is based on the geometry information of parts and systems and enables detailed packaging and kinematic analysis. One key benefit from DMU is that any abnormal stresses in the back, arms, or legs while lifting, moving, or carrying equipment would be instantly highlighted. As a result, access to equipment and removal procedures have been simulated and optimized.

For example, ladder and access platforms have been built into the airframe behind the rear pressure bulkhead to provide rapid access to the stabilizer screw-jack and its associated equipment. The avionics bay has three entrances: via an exterior hatch, through the forward cargo hold, and via a ladder from a hatch in the cockpit floor. The bay is so big that even tall engineers can stand upright. Other emergency equipment resides in a compartment above the cockpit while further built-in platforms provide access to the radome through the nose-wheel bay.

Fulfilling this objective spawned another first for the project: the creation of a computing model to predict the aircraft's reliability/maintainability at EIS. This is currently used to monitor progress in potential R&M improvements, but when critical systems begin to be tested on test-benches and during flight tests, it will enable these predictions to be compared with actual failures and removals. The results then will improve the predictive model prior to the A380's EIS.

The A380 MRBR proposal acceptance by EASA is the first major step in the development of the A380 scheduled maintenance document. It was not the first one, and it will not be the last. To further optimize the maintainability of the A380, it is essential that from day one, the operators collect in appropriate databases the results of the scheduled maintenance tasks to later support the improvement of the MRBR. This will be a continuous process.

As a result, the first technical evaluations of the A380 maintenance program, based on standard aircraft operations, confirm that the initial MRBR targets are met [112]:

- An equivalent A check can be scheduled every 750 flight hours
- An equivalent C check can be scheduled every 24 months/6,000 flight hours
- Structure inspections can be scheduled every 6 and 12 years

This leads to significant maintenance man-hours and maintenance cost savings compared to aircraft with similar operations (e.g., Boeing 747 series). And, last but not least, according to Airbus, through further optimization, even much more sufficiently robust accommodations for A380 check intervals will be achieved through its maintainability optimization methodologies.

11.6 CONCLUSIONS

In this chapter, some core R&M methodologies in Airbus A380 program have been demonstrated. The reliability modeling approach using Markov chain theory has shown that it can enable adequate prediction for the global OR performance of the A380 aircraft. This model is capable of integrating a wide range of useful information, including system and maintenance parameters. Together with the initial design targets of the aircraft, this approach empowers validation and verification for the A380 program in the early design stage, which can significantly reduce the development cycle of the aircraft. The reliability enhancement process, through comprehensive analysis of in-service/history data and engineering concerns, has been shown to be sufficient to generate practical and realistic quantitative analysis of the aircraft reliability and safety. It can be able to provide the reliability prediction with a higher confidence level. This process is also a continuous approach throughout the service life of the aircraft, which as well provides an effective way of engineering experience tracking for any future aircraft (system) development. Last but not least, this paper has shown some major advanced maintainability design and optimization methodologies in the Airbus A380 program, which have occupied a vital role in assuring that the highly demanding reliability targets of A380 can be met throughout its service life.

Since the first flight of the Airbus A380, the approach of its reliability and maintainability methodologies have proven to be successful. Methodologies discussed in this chapter are not only valid in the case of the A380 but, as a matter of fact, they can be used as valuable references in any related area, including new aircraft development, aviation system design, and other industrial applications.

References

[1] Global Market Forecast: The future of flying 2006–2025, Airbus S.A.S., Sia, Lavaur, France; 2006.

[2] http://www.chinadaily.com.cn/china/2013-09/25/content_16994631.htm

[3] COMAC. COMAC's annual forecast report on China's civil aircraft market (2013–2032). Shanghai: Commercial Aircraft Corporation of China; 2013.

[4] Blanchard BS. Logistics engineering and management. 6th ed. Upper Saddle River, NJ: Pearson Prentice Hall; 2004.

[5] Hastings, NAJ, Asset management and maintenance. Brisbane, Queensland: Queensland University of Technology; 2000.

[6] McLoughlin B. Maintenance program enhancements. AERO *magazine*. Shannon Frew; 2006. p. 26–32.

[7] Kinnison HA. Aviation maintenance management. NY: McGraw-Hill; 2004.

[8] Operator/manufacturer scheduled maintenance development. Washington DC: Air Transport Association of America (ATA); 2009.

[9] Pora J, Hinrichsen J. Material and technology development for the A380. In: Proceedings of the 22nd International SAMPE Europe Conference of the Society for the Advancement of Materials and Process Engineering, Paris; 2001.

[10] Hale J. Boeing 787 from the ground up. AERO. Shannon Frew: Boeing Company; 2006. p. 17–20.

[11] Marsh G. Airbus takes on Boeing with composite A350 XWB 2008. Available from: http://www.reinforcedplastics.com/view/1106/airbus-takes-on-boeing-with-composite-a350-xwb/

[12] http://en.wikipedia.org/wiki/Composite_material

[13] Kaw AK, editor. Mechanics of composite materials. 2nd ed. Boca Raton, FL: CRC press; 2005.

[14] Wang B. Composite structural design. Aircraft Design Handbook (Structure Design). Beijing: Aviation Industry Press; 2000. p. 631–685.

[15] Daniel LM, Ishai O. Engineering mechanics of composite materials. New York: Oxford University Press; 1994.

[16] Smith DJ. Reliability maintainability and risk. 8th ed.: Practical methods for engineers including reliability centred maintenance and safety-related systems. Amsterdam, Netherlands: Elsevier; 2011.

[17] Amstadter B. Reliability mathematics fundamental practices procedures, vol. 1. NY: McGraw-Hill Companies; 1971.

[18] Liu Y, Kerre EE. An overview of fuzzy quantifiers (I). Interpretations. Fuzzy Sets Syst 1998;95(1):1–21.

[19] Nowlan FS, Heap H. Reliability-centred maintenance. Springfield, VA: National Technical Information Service, US Department of Commerce; 1978.

[20] Moubray J. Reliability-centered maintenance. 2nd ed. Jordan Hill, Oxford: Butterworth-Heinemann, Linacre House; 1997.

[21] Rausand M. Reliability Centered Maintenance. Reliability Eng System Safe; 1998;60(2):121–32.

[22] Smith AM. Reliability-centred maintenance. New York: McGraw-Hill; 1993.

Reliability Based Aircraft Maintenance Optimization and Applications
http://dx.doi.org/10.1016/B978-0-12-812668-4.00012-5

[23] Backlund F. Conclusion from planning and preparation of RCM implementation. In: The International Conference of Maintenance Societies, Brisbane, Australia; 2002.
[24] Careless J., Cabin security. Avionics Magazine, 2004;28(8):p. 34–8.
[25] Arthasartsri S. Ultra large aircraft (ULA) computerized maintenance management-based reliability-centred maintenance (RCM) methodologies and related airline maintenance program optimization; 2009.
[26] Burchell B. A380: the maintenance-friendly giant? NY: Overhaul & Maintenance; 2005.
[27] Friend CH. Aircraft maintenance management. United Kingdom: Longman Scientific and Technical; 1992.
[28] Hertzberg RW. Deformation and fracture mechanics of engineering materials. 3rd ed. New York: Wiley; 1989. 680.
[29] AMCP706-134: Maintainability guide for design.
[30] MIL-HDBK-472: Maintainability prediction.
[31] GJB/Z 91-97 (FL0112): Maintainability Design Technique Handbook.
[32] MIL-STD-471: Maintainability demonstration.
[33] Australia, S., AS3960: Guide to reliability and maintainability program management; 1990.
[34] Ordnance System Command, U., NAVORD OD 39223: Maintainability Engineering Handbook, Department of Navy; 1970.
[35] Wu H, Zuo HF, Sun W. A study on the accidental damage inspection intervals of aircraft structure based on the improved AHP. Aircraft Des 2008;2008(03):57–61.
[36] Zuo H, et al. Aviation maintenance engineering analysis. Beijing: Science Press; 2011.
[37] Liu M, et al. Research on a case-based decision support system for Aircraft Maintenance Review Board Report. Berlin/Heidelberg: Springer; 2006. pp. 1030–39.
[38] Jing GL, Du WT, Guo YY. Studies on prediction of separation percent in electrodialysis process via BP neural networks and improved BP algorithms. Desalination 2012;291 (April 2):78–93.
[39] Siddique N, Adeli H. Neural networks. Computational intelligence. Hoboken, NJ: John Wiley & Sons; 2013. p. 103–57.
[40] Zhuo JW, Wei YS, Qin J. MATLAB application in mathematical modeling. Beijing: Beihang University Press; 2011.
[41] Rumelhart DE, Hinton GE, Williams RJ. Learning representations by back-propagating errors. Nature 1986;323(9):533–6.
[42] Beale MH, Hagan MT, Demuth HB. Neural Network Toolbox™ User's Guide. MathWorks: Natick; 2012.
[43] Guo QS. System modeling principle and method. Beijing: National Defence Industry Press; 2003.
[44] 737-600/700/800 Maintenance Program Development Policy and Procedures Handbook. Boeing Company: Seattle, Washington, USA; 1996.
[45] Matthews FL, Rawlings RD. Composite materials: engineering and science. FL: CRC Press; 1999. 470.
[46] Bogdanoff J, Kozin F. Probabilistic models of cumulative damage. New York: John Wiley & Sons, Wiley-Interscience; 1985.
[47] Lin KY, Andrey S. Probabilistic approach to damage tolerance design of aircraft composite structures. J Aircraft 2007;44(4):1309–17.
[48] Guidelines and Methods for Conducting the Safety Assessment Process on Civil Airborne Systems and Equipment. Warrendale: SAE International; 1996.
[49] Sirivedin S, et al. Matrix crack propagation criteria for model short-carbon fibre/epoxy composites. Compos Sci Technol 2000;60(15):2835–47.

[50] Chryssanthopoulos MK, Giavotto V, Poggi C. Characterization of manufacturing effects for buckling-sensitive composite cylinders. Compos Manuf 1995;6(2):93–101.

[51] Stig F, Hallström S. A modelling framework for composites containing 3D reinforcement. Compos Struct 2012;94(9):2895–901.

[52] Vesely WE, et al. Fault tree handbook. Washington, DC: Nuclear Regulatory Commission; 1981.

[53] Stamatelatos M, Vesely W. Fault tree handbook with aerospace applications. Washington, DC: NASA Publications; 2002.

[54] Collet J. Some remarks on rare-event approximation. Microelectron Reliab 1997;37(4):687.

[55] Demichela M, et al. On the numerical solution of fault trees. Reliab Eng Syst Safe 2003;82(2):141–7.

[56] Dong Y, Yu D. Estimation of failure probability of oil and gas transmission pipelines by fuzzy fault tree analysis. J Loss Prevent Proc 2005;18:83–8.

[57] Dixit US, Joshi SN, Davim JP. Incorporation of material behavior in modeling of metal forming and machining processes: a review. Mater Design 2011;32(7):3655–70.

[58] Long MW, Narciso JD. Probabilistic design methodology for composite aircraft structures. Washington, DC: Northrop Grumman Commercial Aircraft Division; 1999.

[59] McCarty JE, Johnson RW, Wilson DR. 737 graphite-epoxy horizontal stabilizer certification. In: 23rd Structures, Structural Dynamics and Materials Conference. AIAA, New Orleans, LA; 1982.

[60] Takaki J, et al. CFRP horizontal stabilizer development test program. In: ICCM/9 Composites Properites and Applications. Madrid: University of Zaragoza; 1993. p. 151–58.

[61] Ford T. ATR composite wing. Aircraft Eng Aerospace Technol 1993;65(9/10):22–4.

[62] A380 Policy and Procedure Handbook. Airbus Industry; 2006.

[63] 787 Scheduled Maintenance Requirements Development Policy and Procedures Handbook *(PPH)*. The Boeing Company; 2007. p. 363.

[64] Tropis A., et al. Certification of the Composite Outer Wing of the ATR72. Proceedings of the Institution of Mechanical Engineers, Part G, J Aerospace Eng, 1995. 209(4): p. 327–339.

[65] Federal aviation regulations, Part 25: airworthiness standards: transport category airplanes, F.A.A. (FAA).

[66] Ren H, Steiner T, Wang X. Air-vessel corrosion damage distribution and reliability modeling. J Aircraft 2010;47(6):2115–8.

[67] Company B. 737-800 Structural Repair Manual. Seattle, Washington DC: Boeing Company; 2011.

[68] Xiong H et al. Visual inspection influence factors study of composite structures. In: The 17th National Composite Academic Conference, Beijing; 2012. p. 803–806.

[69] Cary H, Kuen L. A method for reliability assessment of aircraft structures subject to accidental damage. In: 46th AIAA/ASME/ASCE/AHS/ASC Structures, Structural Dynamics and Materials Conference, American Institute of Aeronautics and Astronautics; 2005.

[70] Ushakov A, et al. Probabilistic design of damage tolerant composite aircraft structures. Moscow: Central Aerohydrodynamic Institute (TsAGI); 2002.

[71] Zhou P, Zhao X. MATLAB mathematical modeling and simulation, vol. 1. Beijing, China: National Defence Industry Press; 2009. 384.

[72] Mahadevan S, Dey A. Adaptive Monte Carlo simulation for time-variant reliability analysis of brittle structures. AIAA J 1997;35(2):321–6.

[73] Wang A, Li Q, Chen P. Probabilistic damage tolerance analysis methodology of composite civil aircraft structures. In: The 17th National Composite Academic Conference, Beijing; 2012. 258–264.

[74] Lin KY, Styuart AV. Probabilistic approach to damage tolerance design of aircraft composite structures. J Aircraft 2007;44(4):1309–17.

[75] Verma AK, Ajit S, Karanki DR. Reliability and safety engineering. NY: Springer; 2010. 557.
[76] Armstrong KB, Cole W, Bevan G. Care and repair of advanced composites. London: SAE International; 2005.
[77] Xie J, Lu Y. Study on airworthiness requirements of composite aircraft structure for transport category aircraft in FAA. Procedia Eng 2011;17:270–8.
[78] Chen S. Repair guidance for composite structures. Beijing, China: Aviation Industry Press; 2004. 205.
[79] Ilcewicz L, Cheng L, Hafenricher J, Seaton C. Guidelines for the development of a critical composite maintenance and repair issues awareness course. Washington, DC: DOT/FAA/AR-08/54; 2009.
[80] Andrews JD, Clifton AE II. Fault tree and Markov analysis applied to various design complexities. In: Proceedings of the 18th International System Safety Conference, Texas; 2000.
[81] Misiti M, et al. Wavelet toolbox user's guide. Natick: MathWorks; 1996. p. 626.
[82] Sohn H, et al. A review of structural health monitoring literature: 1996–2001. New Mexico: Los Alamos National Laboratory; 2004.
[83] Farrar CR, Worden K. An introduction to structural health monitoring. Philos T R Soc A 2007;365(1851):303–15.
[84] Defence Standard 00-970/Issue 1. In: Design and Airworthiness Requirements for Service Aircraft. 2007, London, UK: Ministry of Defence.
[85] Aircraft structural integrity program (ASIP)', MIL-STD1530C (USAF). Washington, DC: Department of Defense Standard Practice; 2005.
[86] Kobayashi M, et al. Structural health monitoring of composites using integrated ultrasonic transducers. In: The 15th International Symposium on Smart Structures and Materials and Nondestructive Evaluation and Health Monitoring. International Society for Optics and Photonics; 2008.
[87] Speckmann H, Roesner H. ECNDT 2006—Tu.1.1.1 structural health monitoring: a contribution to the intelligent aircraft structure; 2006.
[88] Andrew KSJ, Daming L, Dragan B. A review on machinery diagnostics and prognostics implementing condition-based maintenance. Mech Syst Signal Process 2006;20:1483–510.
[89] ARP6461—Guidelines for Implementation of Structural Health Monitoring on Fixed Wing Aircraft. Warrendale: SAE International; 2013.
[90] Pintelon LM, Gelders LF. Maintenance management decision making. Eur J Oper Res 1992;58(3):301–17.
[91] Tinga T. Principles of loads and failure mechanisms. Applications in maintenance, reliability and design. Springer Series in Reliability Engineering. London: Springer Verlag; 2013. p. 302.
[92] Ahmadi A, Soderholm P, Kumar U. Reviews and case studies: on aircraft scheduled maintenance program development. J Qual Maint Eng 2010;16(3):229–55.
[93] Stecki JS, Rudov-Clark S, Stecki C. The rise and fall of CBM. Key Engineering Materials, 588; 2013. p. 290–301.
[94] Wenk L. Guidance update on using SHM for continued airworthiness of aero structures. In: Fifth European Workshop on Structural Health Monitoring. Pennsylvania: DEStech Publications; 2010.
[95] Further advanced definition of structural health monitoring (SHM)/addition to MSG-3. ATA MSG-3 SHM Working Group; 2009.
[96] Norris G, Kingsley-Jones M, Learmount D, Phelan M. Europe's giant. Flight International Supplement, 2003.

[97] Arthasartsri S, Ren H. Validation and verification methodologies in A380 aircraft reliability program. In: Reliability, Maintainability and Safety, 2009. ICRMS 2009. IEEE 8th International Conferencen; 2009.

[98] Cutler D. A380 Maintenance Status Report. Airbus World FAST Magazine, 28. p. 14.

[99] Marshall J, Lumbard D, Tanner, G. A comparison of reliability methods and the REMM process, in RAMS 2004. Los Angeles; 2004.

[100] Burger T, Stilp T. A380 Briefing. A380 - Airport Ready, Airbus World FAST Magazine #37; 2005. p. 23–7.

[101] Swink ML, Sandvig JS, Mabert VA. Adding zip to product development: concurrent engineering. Business Horizons 1996;39(2):41–9.

[102] Sharma KJ, Bowonder B. The making of Boeing 777: a case study in concurrent engineering. Manufact Technol Manag 2004;6(4).

[103] Sharma TC, Bazovsky I.Sr. Reliability analysis of large system by Markov techniques. Reliability and Maintainability Symposium; 1993. p. 260–67.

[104] Cocozza-Thevend. Processus stochastiques et fiabilite des systemes. Berlin: Springer-Verlag; 1997.

[105] Hugues, E. Application of Markov processes to predict aircraft operational reliability. In: 3rd European System Engineering Conference, Toulouse; 2002.

[106] Barlow P. Mathematical theory of reliability. Philadelphia: SIAM; 1996.

[107] Marshall J, Newman R. Reliability enhancement methodology and modeling. London: Avionics; 1998.

[108] Marshall J, Jones J. Practical evaluation of the REMM process. In: Reliability and maintainability symposium, IEEE RAMS'06 Annual; 2006.

[109] Babuška I, Nobile F, Tempone R. Reliability of computational science. Numer Methods Partial Differen Eq 2007;23(4):753–84.

[110] Benac C. A380 simulation models. Airbus Standardization of Developments New Media Support Centre [OL]; 2003.

[111] Sweetman B. A 380 flight testing gathers pace. Aircraft Technol Eng Maint 2005;(76):26.

[112] Delmas RB, Broutee R. The A380 maintenance program is born! Airbus World FAST Magazine; July, 2006. p. 11–9.

Index

B

C

D

E

F

G

H

I

K

L

M

Printed and bound by CPI Group (UK) Ltd, Croydon, CR0 4YY

08/05/2025

01864800-0001